It is obvious Mr. Martin has years of practical experience, which becomes very valuable when teaching and explaining conditioning protocols to inexperienced students. I was extremely impressed by the strength training section because I have been a professional strength coach for 22 years. I firmly believe every bit of information he described and prescribed in this section.

—Coach John Philbin, NFL/Olympic Strength Coach

Burgers & Milkshakes

Burgers & Milkshakes

◆

A Pathway Toward Improved Fitness

for the Fitness Enthusiast and Personal Trainer

David B. Martin, CCS

iUniverse, Inc.

New York Lincoln Shanghai

Burgers & Milkshakes
A Pathway Toward Improved Fitness

iUniverse books may be ordered through booksellers or by contacting:

iUniverse
2021 Pine Lake Road, Suite 100
Lincoln, NE 68512
www.iuniverse.com
1-800-Authors (1-800-288-4677)

ISBN-13: 978-0-595-34776-6 (pbk)
ISBN-13: 978-0-595-79515-4 (ebk)
ISBN-10: 0-595-34776-2 (pbk)
ISBN-10: 0-595-79515-3 (ebk)

Printed in the United States of America

Contents

ACKNOWLEDGMENTS

Writing this book was one of the most stressful, time-consuming projects that I have ever completed. What started as a research project for school to graduate ended up being more than I imagined.

I want to thank both my parents for always believing in me and me dreams, even though I aged them prematurely with my low key personality.

I want to thank my wife Lisa who was there when I started this project and kept me focused on the finish.

I want to thank my brother Michael for his words of inspiration and motivation that kept me in full throttle for this adventure.

I want to thank my son Noah, who reminds me each day of the purity and inno-cence of kids, and how much pride he gives me.

Lastly, thanks to all of my friends and clients who made this vision possible.

FOREWORD

I know what you must be thinking: "Oh no, not **another** fitness book." Well, yes…and no. While this certainly is a fitness book, it's not a run-of-the-mill fitness manual. It offers a pathway to improved fitness for both the fitness enthusiast (new or experienced) and the aspiring or certified personal trainer. My knowledge is drawn from educational sources as well as years of hands-on experience working with everyone from the average person in the gym to the elite, Olympic-level athlete on the field.

I have been involved in the health and fitness industry in one way or another for many, many years. Whether I was studying Physical Education and Sports Management in college, working as an Assistant Strength and Conditioning Coach with the NFL's Baltimore Ravens and The United States Olympic Bobsled Team, running my own personal training company, or conducting health and training seminars, I've noticed that no one is immune to the gamut of health and fitness crazes flooding our lives. No one is immune to the search for a health and fitness "quick fix".

Diets seem to cause people the most anguish. There is a constant war over what we should or should not eat. Fearing the frown of disapproval from others, we seek the endless diets. But, we neglect adopting a healthy lifestyle of eating, which will last a lifetime and bode us well. The grapefruit diets, the "48-hour liquid diet", the cabbage-juice diet, and the Hollywood diet—all merely promote yo-yo dieting and weight loss (which is primarily water and muscle) and the inevitable rebound weight gain. With unsound yo-yo diets, no long-term benefits can ever be achieved.

You can't imagine my frustration as a fitness educator, competing against the burgeoning fitness-industry exercise innovators, entrepreneurs, and quick-buck supplement companies who continually try to reinvent the wheel based on their "latest, proven technique." They know that more money can be made if something "new and exciting" comes along. Why do you think there are numerous exercises for each body part, when perhaps 2 or 3 are more than sufficient? The age-old, proven methods of strength training, sensible eating habits, and cardiovascular training are the only prescription for a long and healthy life.

Don't be mistaken: some training concepts have been debunked over time, some have been updated, and others just laughed off the block. Day in and day out, people tend to seek the easiest way to train their bodies for health and longevity. Quite honestly, I'm not going to provide you any "revolutionary" training technique that came to me in the middle of the night.

What has become quite evident as I've looked back on the people I've trained is that they all battle similar areas of frustration. I wish I could wave a magic wand and commit people to living a well-rounded and consistent healthy lifestyle. But my wand is time, commitment, patience, and a sound foundation of education.

I offer a basic book of proven techniques that have worked for many. I will not offer 40 different ways to do an ab exercise when just one would do. I have no intention of reinventing the wheel; just a return to basics.

Similarly, I'm not going to discuss "diets" in detail. Once again, there are basic components to healthy, well-balanced eating that do not inflict more harm than good on the human body.

I endorse moderation when eating, but do not preach sainthood. Quite frankly, my idea of a classic meal is a 1/4-pound flame-cooked burger and a chocolate malted milkshake, finished off with a few Krispy Kreme donuts. I know this sounds excessive, even (dare I say) gluttonous, and goes against what "other" fitness people spout, but I have exceptions: One, I live in the real world, and two, I practice moderation with my diet. I did not say my preferred meal is consumed daily, because it isn't. What I do is exercise daily, eat small balanced nutritional meals 6 or 7 times a day, rest when needed, and also remember to give myself a treat perhaps once a month. Burgers and Milkshakes provides the necessary basic ingredients to assist you with improving your current fitness level, maintaining it, or taking it to a new level with a complete strength and conditioning program.

Adherence to a strict diet, performing hours of cardio each week, performing strength training exercises, and having the added benefit of good genetics, will, without question, work. But rare are the few who have the time, money, dedication, and motivation to achieve results in this manner. The possible frustration of never getting "the look" quickly is further compounded by a monthly slew of fitness magazines with pages filled with God-like, chiseled, and bronzed models. In addition, at least once a year a new fitness gadget surfaces, claiming the delivery of rock-hard abs in just six minutes a day! For most people, the sales pitch will work; however, getting the longed-for six-pack abs using the latest miracle gadget will not.

Burgers and Milkshakes offers you no quick miracles. YOU will become healthier and more fit by taking action steps now: first by reading, then by implementing and incorporating fitness into your life.

INTRODUCTION

Magazines too often claim that you can follow the latest celebrity workout program and have an amazing, lean body. Here's the catch: You need a spare 2-3 hours a day to train, about $200-$300 a day for a trainer, minimal negative stress, and great genetics. Elite, professional, and Olympic athletes are paid to train, maintain an optimal physical condition, and perform at peak condition. That's their job. However, for the average person who puts in a 9-hour workday (and contends with daily traffic jams and the demands of family life), spending several hours a day working out is just not feasible.

But don't be discouraged, for I offer a solution: If you are seeking realistic, time-efficient, and effective fitness programs, then congratulate yourself for selecting this book.

By reading and following the prescriptions inside *Burgers and Milkshakes,* you will make real gains in your pursuit of a healthier life.

Eating healthy, balanced meals (fruits, vegetables, whole grains, and lean low-fat protein) and getting daily exercise for both the heart and muscles will greatly improve your overall health and lower your body fat. Yet occasional indulgences such as burgers and milkshakes never killed anyone, or added pounds. Occasional treats will preserve your sanity.

Burgers and Milkshakes delivers useful information that is easy to read, understand, and incorporate into your lifestyle. There are no celebrities endorsing the "newest" diet. I offer no magic pills. Nor am I selling you the latest fitness gadget. What I do offer are facts supported by scientific research. Unlike celebrities or athletes, most people cannot afford a daily personal trainer and a dietician to select and prepare healthy foods. *Burgers and Milkshakes* is for real people who want real results.

Each chapter provides information that will ensure your success with a fitness lifestyle change, along with scientific research to support the chapter's topic. I also strongly recommend further fitness education *via* taking college courses, obtaining fitness certifications, and attending fitness workshops and lectures. Despite what the fitness magazine industry would like you to believe, it is not the authority on scientific research.

Burgers and Milkshakes is unique with its information for the personal trainer and the general fitness enthusiast. If you are a baby boomer looking to tap into the fountain of youth, a tennis player trying to improve your game, or a young athlete trying to make the talent scouts take notice, this book will give you the tools to reach your fitness goals.

Burgers and Milkshakes has the aspiring personal trainer in mind. As a trainer, you will have a greater impact on your business by applying the information. However, if you are a regular person who seeks improved fitness, when you see the word "client", substitute your name and follow along. Additionally, carry this book with you to help you stay focused, give you ideas to enhance your program, and assist you in designing a wide variety of fitness programs.

ACHIEVING BALANCED HEALTH

An effective strength and conditioning program is a balanced one; it conditions the whole body, allowing you to function more efficiently and reducing the potential for injury. The integral components of a balanced conditioning program are:

- Flexibility
- Muscular strength
- Cardiovascular training
- Nutrition
- Rest/recovery
- Specific sport training
- Variety
- Stress management

Despite what popular fitness magazines say, and what the dietary supplement industry would lead you to believe, there are no magic pills or potions for acquiring a fit and *healthy* body. The only way results will be achieved is with a personal commitment to positive lifestyles changes and to a balanced fitness and conditioning program.

Flexibility and stretching

Flexibility is probably the most neglected area of strength and conditioning programs. People tend to underestimate the benefits of a stretching program. Although stretching should be done daily, it rarely is. Often a lack of time and perseverance (stretching should never be rushed!) causes most people to short-change this aspect of an exercise program. But stretching reduces potential injuries and alleviates stress on the body. Stretching on a regular basis can ease tension and improve all body movements, resulting in less joint pain, increased range of motion, reduced muscle tension, increased mental readiness, and increased blood circulation (NSPA, 1997).

A warm-up period of 5-10 minutes of light aerobic activity prior to stretching is recommended. This warm-up and stretching session will increase blood circulation and body temperature. In addition, it mentally prepares you for the upcoming training session. Never stretch cold muscles; cooler muscle-tendon units are stiffer and more easily injured than warm muscles (NSPA, 1997); such injuries may be micro-tissue tears. In addition, avoid stretching the muscle too far. Any time a muscle is stretched too far, a nerve reflex (called a muscle or neuromuscular spindle), responds by signaling the muscle to contract. This neurophysiological reaction is one way the body protects itself from injury. While the muscle spindle indicates muscle length, the golgi tendon organ (a proprioceptive receptor located within the tendons found on each end of a muscle) indicates muscle tension. The golgi tension organ responds to increased muscle tension or contraction as exerted on the tendon by inhibiting further muscle contraction. When muscle contraction is excessive, the golgi tendon organ protects against muscle damage.

The most popular stretch is the static stretch. The static stretch slowly lengthens the muscle to a stretched position, or just to the point of gentle tension. The muscle is held for a fixed period of time (usually 20-40 seconds) (NSPA, 1997). The American College of Sports Medicine (ACSM) recommends 3-5 repetitions for each stretching exercise (ACSM, 1995). Remember to breathe throughout the stretch. Sometimes an entire session devoted to stretching may be necessary.

Muscular strength

Muscular strength is the muscle's ability to exert an external force against resistance during a single maximum contraction. Strength training with proper progressive overload will help you improve your strength. To train for muscular strength, perform 6-12 repetitions with a weight heavy enough to make the last repetition difficult to lift with proper form (no jerking or "cheating").

Muscular endurance

Muscular endurance is the ability to persist with physical activity (for example, by performing repetitive muscular contractions against some resistance) for an extended period of time. Repetition ranges from 15-20 help to develop increased muscular endurance. If you are new to strength training, it is recommended that you first develop muscular endurance in order to establish a strength base. As muscular strength increases, so will muscular endurance. Without muscular strength and endurance it won't really matter if you are trying to train for cardiovascular endurance. Take running as an example: If your leg muscles lack the muscular strength and endurance to help you run for more than a few minutes, you certainly will not be able to run long enough to develop any sort of cardiovascular endurance.

Cardiovascular fitness

While muscular endurance depends on the efficiency of the skeletal muscles and the nerves that control them, cardiovascular endurance depends upon the efficiency of the heart muscle, the circulatory system, and the respiratory system, and their combined ability to supply the body's muscles with adequate amounts of oxygen. Daily aerobic and anaerobic training will increase cardiovascular capacity as well as provide a myriad of other benefits. According to ACSM, at least 30 minutes of enjoyable activity, using all major muscles, performed 4 times per week, is enough to promote cardiovascular benefits (ACSM, 1995).

Nutrition

Most people underestimate the important role that nutrition plays in exercise and health. Without an adequate diet—eating too much fat, too many carbs, and not enough protein—all the exercise in the world will not help to achieve a healthy body and pleasing physique.

The dietary supplement industry adds to the confusion over what are the right foods to eat. Annually, Americans spend billions on diet pills and nutritional supplements. One reason is that most Americans' average days are in full throttle and good nutrition tends to suffer. The supplement industry has seized the opportunity to take advantage of people on the go, targeting those who seek alternatives to fast-food jaunts. It has cleverly convinced us that their products can take the place of proper, well-balanced meals. While replacement meals are a better alternative to fatty fast foods, proper, nutritious foods best enhance health and the quality of life. A balanced diet—one consisting of complex carbs, lean proteins, and a limited intake of saturated fat—improves both one's internal health and physical appearance.

Begin logging your daily food intake in a journal to make you more aware of what you eat. If you are working with a trainer, you may find that they ask to review your food journals, and then discuss them with you. It is not your trainer's job to reprimand you if you happen to slip from the intended diet, as this happens to the best of us. Nor is it your trainer's job to embarrass you. If you feel that you are being treated in such a manner, search for a new trainer. You must be honest when entering your food intake. Your trainer can only help you if you log in ALL foods eaten. This is the only way you will have an honest discussion about your diet, and receive honest suggestions for improvement. For example, you may be unaware that you are eating too few or too many calories. You may not realize that you are not eating the proper amount of wholesome foods such as whole grains, vegetables, fruits, and lean meats. Perhaps you are eating at the wrong time of the day. Only by keeping an accurate daily food journal will you find problem areas, and hopefully craft solutions to eliminate them.

Proper hydration is imperative; sip at least 64 ounces of water daily. For in-depth details on nutrition, contact a Registered Dietician.

Another method for measuring your performance is by having your body fat percentage assessed. This is where a trainer or staff member at a gym facility comes in handy. Charges may differ for performing this test; sometimes it may be free. Excessive weight loss is not the key to lowering your body fat percentage; the maximum recommended weight loss should be 2 pounds per week. Many people find that they are actually gaining weight and consequently become disappointed in their progress. Realize that this particular weight gain is actually a result of the strength training program. While the scale may indicate that you are gaining weight, you will experience a leaner appearance due to increased muscle mass and reduced body fat. A sign of a good trainer, if you are using one, is that he or she educates you on the fact that muscle weighs more than fat, and that scale weight and actual body fat-to-muscle ratios are two different measures. Do not be ruled by your scales; rather, start a new habit of looking in the mirror more often and seeing the results for yourself.

Rest and recovery

In the quest for healthier bodies, people often tend to overdo it. Everyone knows the over-enthusiastic trainee who sets outs doing cardio 6 days a week, strength training 4 days a week, and additional aerobic classes for insurance. These people inevitably crash and burn. If you find yourself gimping around feeling pain and dreading your next exercise session, then you've overdone it! **In order for the**

body to heal and grow, it needs rest. If you do not allow enough time between your training sessions for proper recovery, you are setting yourself up for potential injuries. About 2 hard intense training days a week is adequate for improving your overall heath and fitness level. Train hard, go home, enjoy life, rest, and then come back to the gym ready for more.

Sport-specific training

Sport-specific training is a form of exercise targeted to particular muscle groups, energy pathways, and movement patterns that are involved in the performance of a specific activity. Basic patterns of movement such as lateral, vertical, forward, backward, rotation, or combinations thereof are associated with most sports. If you are an athlete, remember to incorporate a well-planned strength and conditioning program that is tailored to your particular sport. The program should involve patterns of movement, which mimic those used on the field. Additional time in the weight room doing more sets only wastes energy and valuable rest time, and is not the answer. Practice as if you were playing your sport.

Variety

Maintaining enthusiasm and interest is a major challenge during fitness training programs. Add variety to your program by using different equipment, changing the order of exercises, and altering the sets, intensity, recovery, repetitions, and your training environment. Avoid the ordinary; think of safe ways to achieve results. Creativity will ensure satisfaction.

Stress management

Stress management is a term for a collection of techniques used for managing the physiological and mental stressors in our lives. Used properly, stress-reducing techniques will help keep the body and mind in balance. If you know you are stressed, you're not alone; most of us live extremely stressful lifestyles, which contribute both positively and negatively to our health. Identify negative stressors in your own life and then explore possible stress-reducing techniques, such as relaxation tapes, yoga, meditation, [therapeutic, to distinguish from "other" massages] massage, and exercise. All are great relievers of stress. Determine what works best for you and then incorporate such stress management skills into your daily life.

A CRASH COURSE IN ANATOMY

An important aspect of personal training and enhancing your level of fitness rests on your understanding of basic anatomy and physiology and its relationship to strength training and conditioning. Basic comprehension of how a muscle contracts and relaxes, its energy requirements, where muscles and organs are located, and what mechanics are involved in muscle movement will give you a greater appreciation for your body. This knowledge provides you with pertinent information to design and execute an individualized program that is safe and effective.

Skeletal muscle anatomy

Essentially, muscle physiology includes bones, tendons, ligaments, and muscles.

Tendons attach muscle to bone, and concentrate a pulling force in a limited area. Tendons are capable of withstanding great force. Ligaments are bands of connective tissue that connect bone to bone. The muscle end attached to the least movable part is called the origin. The other end, the movable part, is called the insertion. Each muscle is made up of divisions, and each division is wrapped in connective tissue known as fascia. The outer layer of fascia is called epimysium. The connective tissue, which encloses the fascicle, is called the perimysium. The fascicle is a group of muscle fibers bundled together. The endomysium, which is a thin tissue, surrounds each single muscle fiber. The sarcolema, located beneath the endomysium, surrounds the sarcoplasm.

Muscle fibers and filaments

Muscle fibers consist of thin strands called myofibrils, which are cylindrical structures that run longitudinally and parallel to each other through the muscle fiber. Myofibrils consist of smaller structures called myofilaments, which in turn consist of two contractile proteins: myosin (thick) and actin (thin). Cross bridges, which provide the functional connection between the two filaments for muscle contraction, protrude from the myosin heads. Myofilaments do not run the length of the muscle fiber, but instead are arranged in compartments called sarcomeres, which are the contractile units of the muscle. Narrow zones of dense materials called 'Z lines' separate sarcomeres from each other, and this is where the thin filaments attach. Located within the sarcomere are dense striations. The dark dense band, the 'A band', represents the length of myofilaments. The overlapping of both thick and thin myofilaments causes the darkness of the 'A band.' The lighter, less dense bands, the 'I bands' consist of thin myofilaments only. A region in the middle section of an uncontrolled muscle is called the 'H zone,' and contains thick myofilaments only.

Muscle contraction

The most widely accepted explanation for muscle contraction is the sliding filament theory. It states that during muscle contraction, the actin myofilaments slide past the thick myosin myofilaments to bring the 'Z lines' closer together. It is the simultaneous contraction of the sarcomeres that leads to contraction and shortening of the muscle fibers, which produces a contraction of the muscle as a whole.

When a muscle is stimulated to contract, the myosin cross bridges of the thick myofilaments attach to sites on the actin of the thin myofilament. This acts to pull the actin filaments closer. The cross bridges bond to several successive sites along the actin as the two filaments slide over each other. Though the sarcomere shortens during contraction, the thin and thick myofilaments do not alter. As the thick and thin myofilaments move past one another, the 'A band' stays the same, the 'H zone' narrows and even disappears, and the 'I band' becomes smaller.

Principles of neuromuscular control

In muscle physiology, the "all-or-none principle" states that all muscle fibers associated with a particular motor neuron contract to their fullest extent when stimulated, or not at all. A motor neuron is responsible for transmitting signals to muscle tissue. A motor neuron and all the muscle fibers it controls are called a motor unit.

The "size principle" states that selection of motor neuron size and its associated muscle fiber types follow an order from smallest to largest.

Muscle fiber types

There are two major types of muscle fibers: fast twitch and slow twitch. Fast twitch fibers (type I) are used in maximal force contractions and are low in myoglobin. They can only produce force for short periods of time and are principally used in anaerobic metabolism. Activities such as sprinting, football, and power lifting utilize these muscle fibers. Slow twitch fibers (type II), on the other hand, have slow contractile speeds and more mitochondria and myoglobin. They are used in activities requiring aerobic metabolism, such as distance running and endurance activities.

Genetic variables

An individual's physiological response to the components of a well-rounded conditioning program is determined largely by genetic attributes. No two people are alike and no two people respond to exercise in exactly the same way. Some may react similarly to the same training regime, but their responses will not be exactly alike. This, again, is due to the genetic makeup of the individual. I have yet to read in any of the fitness magazines how genetics play a crucial role in muscle development. If you look at these magazines each month at the news stand, they have similar headlines: "Lose fat in 4 weeks, Get Huge Biceps in a Few weeks, Foods that Burn fat etc…" becoming fit, strong and healthy requires self discipline, commitment and great plan. According to Riley (1998), the following genetic variables affect individual responses:

Muscle length: An individual possessing relatively long muscles has a greater potential for muscle growth and strength than an individual with shorter muscles. A muscle cannot develop wider than the length of the muscle belly.

Limb length: The limb length alters the angle of forced output. A muscle attached to short limbs can lift more weight due to mechanical advantage. Individuals with shorter arms, legs, and better insertion points (the manner or place of attachment of a muscle to the bone that it moves) have a greater biomechanical advantage over individuals with longer limbs.

Insertion point: The point of muscle insertion is significant in strength development. If located further away from where the joint of a muscle tendon inserts, there is a greater mechanical advantage.

Somatotype: The body structure can be classified into three basic types according to generalized characteristics: endomorph (fleshy), mesomorph (muscular), and ectomorph (linear). Most of us share simoilarities with two of these body types.

DEVELOPING YOUR STRENGTH AND CONDITIONING PROGRAM

Your goal should be to improve your overall health and fitness while keeping injuries to a minimum. A structured approach ensures that a quality program will be easily developed and administered. You may need the assistance of a trainer or other health and fitness professionals to help you achieve these goals and help plan an individualized program. There are four steps in the planning process:

Step 1: Establish your needs and goals

Consider what it is about your health and physique that you are looking to improve. There are four steps in the planning process that can assist you towards your fitness goals:

Step 1: Orientation

Meet with an educated (preferably earning a degree in the health, physical education), certified personal trainer to discuss your goals, schedule, and medical issues prior to creating and implementing an individualized program. More importantly, you need to learn how to to exercise safely and effectively. Once such a program is established, you should review your goals regularly.

You should reacquaint yourself with your goals during each training session. Your success and sense of achievement will be determined by your ability to meet your predetermined training goals.

Step 2: Assessments

Determining your current fitness level is imperative prior to beginning a program. A comprehensive physical assessment should measure body fat composition, flexibility, muscular strength, muscular endurance, and blood pressure. The results will provide you with key information about your body. The American College of Sports Medicine Table of Guidelines for Exercise Testing and Participation states, "It is not necessary for asymptomatic, healthy men under age 40, and women under age 50, with less than two coronary artery disease risk factors (smoking and obesity), to have a medical examination prior to initiating a program of exercise."

Step 3: Testing

Once an individualized exercise program has been designed for you and is under way, testing is the only way to ensure the program is meeting all of your stated goals and objectives. Review your test results at least twice yearly with or without a trainer. This will keep your motivation high, especially as you begin to see for yourself the improvements being made to your overall health. Maintain periodic testing, as it will identify the need to make modifications to your current program.

Step 4: Education

Education about fitness, the need for a healthy lifestyle, and how to perform exercises correctly is vital to the success of your program. You will develop more confidence with increased knowledge of how the body responds to exercise. Practice continued education. Make the educational process a part of your training program, and seek new information in easily understandable terms. Frequently review old and new information to ensure your retention. As you learn more, you should notice an increased feeling of enjoyment with your exercise program, especially as you begin to apply the learned knowledge, appreciate your body's abilities, and, hopefully, begin to achieve your goals.

GETTING STARTED—FITNESS ASSESSMENTS

Testing and assessment are important for measuring your progress over time. If you are under the guidance of a personal trainer, you should request this protocol.

Testing should be administered a minimum of twice yearly, but to maximize the effectiveness of the overall program, quarterly testing is advised. Continuous fitness testing and assessment is the only way you can see how you are progressing under the current conditioning program. Your personal trainer should discuss the fitness testing and assessment procedures with you to ensure that you understand what you have consented to and what is being tested. During testing, be sure to report any unusual feelings or experiences of discomfort. Your trainer should explain the benefits of the fitness testing, should solicit information or feedback from you, and should satisfactorily answer all your questions or concerns before testing begins.

The tests' results assist a personal trainer with the evaluation of the quality and effectiveness of your current program. In addition, these same test results can provide you with useful information that should be used as a tool to motivate you towards continued progress.

Regardless of whether you consider yourself to be an athlete or an average Joe wishing to improve your health, administered physical assessments and the evaluation of the results is great way to check for improvements. An exceptional software package to help determine one's functional age can be purchased from Fitness in Today's Times, Inc. (301-540-3488). FITT's Functional Age Assessment package provides a comparison of measured functional age to chronological age. Functional age is the age at which your body functions (flexibility, aerobic, conditioning, strength, *etc.*). Based on the results collected, recommendations are offered.

Types of testing

The following tests are normally administered under the guidance of a personal trainer or health and fitness professional. The explanatory procedures are intended to aid those personnel. For norms for each of the tests, refer to Appendix A: Physical Fitness Profile Evaluation. Here are a few tests:

- Aerobic capacity
- Muscular endurance
- Flexibility
- Body fat composition
- Blood pressure

Aerobic capacity

Three-minute step test

The step test measures cardiovascular fitness and can assess the client's present level of fitness.

Equipment:

- 12-inch high bench
- Metronome set at 96 bpm (up, up, down, down to each click), 24 steps/minute
- Timer
- Stethoscope (to count recovery heart rate
- Procedure:

First, demonstrate the step test for the client. Do not allow the client to practice, as this will affect his or her heart rate and create errors in the results. Instruct the client to begin stepping by alternating feet in time with the first click of the metronome. During the test, inform the client how much time is left. Once three minutes has passed, instruct the client to immediately take a seat so that you can begin a one-minute heart rate assessment.

Immediately place the stethoscope on the client's chest, find the heartbeat, and start the timer. Stop the timer when client's heart rate has returned to less than 90 beats per minute. Write down the results.

Muscular endurance

One-minute sit-up test

The one-minute sit-up test is a fair representation of general muscular endurance. This test measures one of the most important muscle groups in prevention of lower back pain: the abdominals.

Equipment:

- Stopwatch
- Exercise mat

Procedure:

First, demonstrate the correct way to perform a sit-up (or crunch). The elbows should touch the thighs as the client comes up to a lifted position. After each movement, the client should return to the supine position before going up again. Do not allow your client to hyper-extend the back.

Instruct the client to assume a supine position on the floor, to bend knees at right angles, and to position heels approximately 18 inches from buttocks. You will need the assistance of a partner to firmly hold your client's ankles. Instruct the client to perform as many correct crunches as possible within a one-minute period. Make sure the client is breathing continuously and easily during the exercise. Upon completion, write down the number of correct repetitions performed.

Flexibility

Sit and reach test

The sit and reach is a flexibility test that provides an overall assessment of trunk and hamstring flexibility. Lower back pain may occur when the hips, lower back, and hamstrings are not flexible.

Equipment:

- Either a trunk flexion instrument or a yardstick and 12 inches of tape

Procedure:

Instruct the client to warm up prior to performing this test. If using the yardstick, place it on the floor. At the 15-inch mark, place a 12-inch long piece of tape at right angles to the stick. Have the client remove his or her shoes and then sit on the floor with legs outstretched in front and positioned on either side of the yardstick. The heels should be lined up with the ends of the tape. The zero mark of the yardstick should be nearest the client. The client's fingertips and hands should be placed on top of one another. Have the client exhale, and slowly lean forward while gradually dropping the head and extending the hands as far possible. Knees should be kept straight throughout the movement, and the stretch should be held for a moment. The distance is calculated measuring the furthest point that the fingertips reached after 3 slow reach attempts. Be sure your client does not bounce into the reach.

Body Composition

Body fat composition is the ratio of fat to lean body mass. The body requires some fat for maintenance and normal bodily functions; fat serves to insulate the body, protect and insulate internal organs, provide the body with energy, and transport fat-soluble vitamins (A, D, E, K).

Body fat skinfold test

The skinfold method, which uses a calibrated skinfold caliper, provides reasonable estimates of fat and fat free mass. Performed properly, skinfold measurement offers more precision than most other available techniques except for underwater weighing (McArdle, 1996).

Equipment:

- Calibrated skinfold calipers

Procedure:

Always take measurements on dry skin and from the right side of the body. Skin folds sites. Using the index finger and thumb, grasp a skinfold of 1/4-1/2 inch deep and position the caliper 1/4-1/2 inch above the site. The caliper head dial should face upward with the head and tips perpendicular to the direction of the skinfold. Position the caliper heads no more than 1/4-1/2 inch deep, then release the grip gently. Read caliper within 1-2 seconds and record, making sure to obtain two readings. The measurements at each site should not differ by more than 1 or 2 mm.

Skinfold measuring sites:

Figure 1

The seven skinfold sites:

- Triceps: A posterior measurement; midway between the acromion process and olecranon process. Vertical pinch.

- Thigh: An anterior measurement; midway between inguinal fold and superior aspect of patella. Vertical pinch.

- Suprailium: A lateral measurement; one inch above iliac crest on the mid-axillary line. Diagonal pinch.

- Chest: An anterior measurement; midway between the anterior axillary line and the nipple. Diagonal pinch.

- Abdomen: An anterior measurement; taken one inch laterally from umbillicus. Vertical pinch.

- Axilla: A lateral measurement; level with xiphoid process of the sternum at the mid-axillary line. Vertical pinch.

- Subscapular: A posterior measurement; one inch below the inferior border of the scapula adjacent to the vertical border. Diagonal pinch.

Source: NSPA, 1997

Ranges of Healthy Body Fat Percentages

Males	6–17%
Females	14–24%

Classification of Body Fat Percentage

	Male	Female
Essential fat	2–4%	10–12%
Endurance	6–8%	14–16%
Conditioned	10–13%	17–20%
Average	14–17%	21–24%
Borderline obesity	18–22%	25–29%
Obese	> 23%	> 30%

Blood pressure

Blood pressure is the term for the pressure in the body's blood vessels resulting from cardiac output and the body's peripheral resistance to blood flow. The pressure in the brachial artery during the contraction of the heart is the point that blood pressure reaches it highest value. This value is called the systolic blood pressure. The second value measured is the diastolic pressure. Diastole refers to the pressure in the artery when the heart relaxes between beats. A normal resting blood pressure (systolic/diastolic) is less than 140/90 mm Hg. Hypertension is defined as a reading greater than 160/95 mm Hg. Hypertension is the most common cardiovascular disease in humans. If the reading is greater than or equal to 140/90, refer the client to a physician for a medical examination and clearance before beginning an exercise program.

Exercise training may be useful in the management of mild hypertension, particularly in the young and middle aged. The majority of studies that report a favorable effect of exercise on blood pressure have used training programs involving moderate intensity exercise (such as aerobics, jogging, or bicycling) several times per week. Higher intensity or more frequent exercise does not appear to offer any advantage and may even be less effective at lowering blood pressure. An exercise program with moderate weight training also appears to be safe and beneficial for persons with mild hypertension (ACSM, 1995).

Blood pressure test
Equipment:

- Sphygmomanometer
- Stethoscope

Procedure:

Readings should be taken from the left arm (due to the proximity to the heart). Two readings should be taken about one minute apart. The client should be seated with arm relaxed at heart level before beginning. Palpate the brachial artery, located medially near the elbow joint. Wrap the cuff snugly around the upper arm, one inch above the elbow joint. Place the stethoscope firmly over the brachial artery. Inflate the cuff to 160 mm Hg, and then slowly deflate at a rate of 2 mm Hg per second. If systolic pressure is heard immediately at 160 mm Hg, repeat the reading and inflate the cuff beyond 160 mm Hg until silence is heard prior to the first occurrence of sound.

The first reading is of the systolic pressure, and is taken at the onset of a sharp, tapping sound. This sound is created as the cuff deflates and blood flow resumes.

The second reading is of the diastolic pressure, and is taken when the sound disappears.

Blood Pressure Readings

Average	120/80 mm Hg
Borderline	140/90 mm Hg
Mild to moderate hypertension	150/95 mm Hg
Moderate to severe hypertension	160/110 mm Hg
Uncontrolled hypertension	170/110 mm Hg

PERFECT FORM STRENGTH TRAINING

How are defined muscles developed? Most people think that muscle definition is the result of performing an exorbitant amount of repetitions. That couldn't be further from the truth: It is not the amount of reps that determines how a muscle develops, *it is how the muscle responds to the stimulus*. Genetics, proper nutrition, intensity, volume of exercise, and rest will stimulate muscular development, not just repetitions alone. A muscle will only adapt to the stress to which it is exposed. Dr. Ellington Darden, former Director of Research for Nautilus Sports Medical Industries and author of numerous exercise-related books, states, **"The building of strength is proportionate to the intensity of exercise.** The higher the intensity, the better muscles are stimulated. Performing an exercise to the point of momentary muscular failure assures that you've trained to maximum intensity."

Repetitions

How many repetitions per set should be completed? Developing size and strength is not a result of quantity of reps, but *how each rep is performed, and the quality of the rep*. The manner in which the repetitions are performed is the foundation of any strength program, and when they are performed properly, they will promote maximum muscular strength.

The number of repetitions performed to fatigue is an important consideration when designing a strength training regimen. The greatest strength gains result from resistance yielding 4-6 repetitions. Increasing the number of repetitions to 12-20 and decreasing the relative amount of resistance will favor increases in muscle endurance (Feingenbaum, 1997). Furthermore, a given percentage of one repetition maximum will not always elicit the same number of repetitions when performing different lifts.

Muscle movements

Depending on the starting point of the repetition, the first phase of movement is termed concentric movement.

A concentric movement is when the muscle is shortened; it should last approximately 2-3 seconds. Once the weight is raised, there should be a momentary pause in the fully contracted position. Squeeze and "feel" the muscle without locking the joint.

The upward movement while performing a bicep curl is an example of the concentric phase of the repetition. When performing a bicep curl, stop when the muscle is fully flexed. A bounce due to forced momentum from lifting quickly indicates that the muscle is performing little or no work through that range. For maximum effectiveness in developing bigger and stronger muscles, lift the weight slowly.

To ensure that no momentum was used with the lift, all movement must be paused before the lowering phase is begun (an eccentric movement).

In contrast to concentric contractions, eccentric movements are contractions in which the muscle exerts force while it lengthens. The eccentric contraction should be controlled during each rep. **Due to gravity, it is easier to lower a weight; therefore, the eccentric movement should be purposely slower than the concentric.** Take approximately 3-4 seconds to lower the movement. During the eccentric phase, the muscle is approximately 20-40% stronger than during the concentric phase. In fact, some studies have shown that the greatest benefits can be gained during the eccentric portion of the movement. More muscle fibers can be recruited when the rep speed is slower, thus increasing the intensity of the exercise.

Quality repetitions

Each rep should look identical in movement and quality, irrespective of whether it is the first, second, or tenth rep. To emphasize this point, imagine that you have taken a photo of each of 10 reps. If all reps have been performed equally, you should not be able to arrange the photos in the order they were performed, then either you have not yet mastered the correct form for this exercise, or the weight is too heavy for you and the load needs to be lightened. Ideally, there should be no difference between the first and last rep.

Unfortunately, it is quite common for people to rush through their reps with no heed to quality. We have all seen this in the gym: the big guy on the bench, raising the 500-lb bar using his hips, back, and legs to assist in the lift. Yes, he got the weight up, but at great risk of injury and not using all the chest muscles (muscle specificity), because many other large muscle groups were helping with momentum.

Train yourself to concentrate on each rep equally, lifting in a controlled manner, as if each rep were the first one performed. When recording your regimen, only record the **correct** number of repetitions lifted. Do not record assisted lifts. For example, if someone assists you with the last two reps out of 15, you should only record 13 complete reps. Performing identical repetitions at a specific prescribed speed will ensure accountability by keeping the variables the same.

An important note: **do not lock the joint**. Locking the joint at the peak of contraction will place undue force on the joint, potentially causing injury, while reducing the tension on the muscles.

Rep speed

When momentum is neutralized by means of isokinetic strength testing equipment, muscles always produce more force (Westcott, 1987). Additionally, after the initial explosive movement, the muscles throughout the remaining range of motion display little or no resistance. The weight almost moves by itself. Maximum muscle force invariably decreases as the speed of movement increases.

To maximize strength and reduce the risk of injury, the weight should be raised at a speed that forces the muscle to perform all the work, with no bouncing, jerking, or sudden movements. Most people perform repetitions too quickly, averaging 1-2 seconds—a speed which greatly reduces muscle fiber recruitment due to gravity and momentum assistance. However, performing one rep in 5-7 seconds will place more demand on the muscle and recruit more fibers (Westcott, 1987).

The advantages of slow repetitions are longer periods of muscle tension, higher levels of muscle force, lower rates of momentum, and less risk of injury. By creating and maintaining tension in the muscle groups, you can force the muscles in question to do more work per repetition. From a metabolic standpoint, this increased work will heighten the intensity of any given exercise and produce a stimulus for proper overload (Mannie, 1997). **Most importantly, slow training produces better strength training results because of the thoroughness of the muscle's involvement** (Darden, 1990). Muscles are recruited in order according to the intensity or force requirements, rather than speed of the movement. Lifting the weight slowly without momentum and recruiting as many muscle fibers as possible achieve greater strength gains.

The goal of the lift is to perform a prescribed number of repetitions and achieve momentary muscular failure (MMF) on the last repetition. Why is this important? It allows the muscle to recruit all the fibers in the muscle group, allowing the muscle to become stronger more quickly. Muscular failure is the point at which the muscles can no longer function.

Always start with a weight that you can lift for 12 repetitions; typically approximately 75% of your maximum strength capacity. If the weight is too light, and 12 or more repetitions are easily completed, the resistance may be too light. Add about 20% more weight; rest a minute, then lift again. If you cannot perform at least eight repetitions, the resistance may be too heavy, and the weight should be lowered.

Injury prevention

Injury prevention is the top priority for any exerciser. Admittedly, people can get huge by using momentum and by throwing and jerking weights, but there is a higher potential for injury to occur.

Correct form execution while lifting weights is critical for increasing strength and reducing the risk of injury. It is important to minimize momentum and bouncing movements when raising and lowering the weight. Exercises involving dynamic (fast) movements are not only unsafe, but they reduce the maximal muscle fiber recruitment. The greater number of fibers recruited, the greater the strength and size potential of the muscle.

Set number

How many sets should you perform? The number of sets to be completed is an important consideration for the strength program. Studies conducted by West-cott (1996) concluded that one set per exercise at maximal effort is sufficient to promote near-optimum strength development. In his study, 38 subjects were separated in to three groups. Each group was put on a 14-week training period and instructed to perform one, two, or three sets of lower body exercise. The group that performed one set showed a 14.5% improvement, whereas the group that performed three sets showed a 15.5% improvement. **So, if your goal is to increase strength in a safe but timely way, one set is your best option.**

One set versus multiple sets

While one advantage of performing one set per exercise is increased strength in a shorter period of time, another is that this training method allows a person to be more focused and work out with a greater intensity due to the decrease in additional demands. Another advantage is a reduced risk of injury through overuse. Finally, performing one set improves the quality of the repetition by allowing you to use a slow, determined, and controlled technique.

Multiple sets

Many people prefer multiple sets. While multiple sets do produce big and strong muscles, they have little to no advantage over using the one-set method. It is purely a matter of preference. One reason for preferring to perform multiple sets may be that some people like the associated "pump"—the rush of blood to the muscle. If you want to feel the muscle swell to maximal potential, perform each set to absolute failure, then move immediately to the next exercise for the duration of the session. Performing multiple sets for each exercise can monopolize time. The extra time gained from the one-set method can be devoted to fat-burning activities, stretching exercises, or rest. Generally, 60 seconds is sufficient recovery time between sets or exercises.

Not surprisingly, there are numerous studies producing empirical evidence that performing one, three, or more sets produce the same results (Mannie, 1997). For more scientific research on the number of sets refer to Appendix B.

High-Intensity Training

There is a way for you to get stronger and larger muscles in less than 30 minutes: For intermediate and advanced persons, each set should be taken to Momentary Muscular Failure (MMF). As discussed earlier, MMF occurs when another repetition using proper form cannot be performed. To develop strength and muscle gains, continue exercising to the point where the completion of another rep is impossible. Admittedly, this is incredibly hard work. But if you are going to strength train, go all out. Push yourself beyond the comfort zone. Your mind may want to give up before your muscles do. If you stop before MMF, additional fibers will not be recruited. Any muscle fiber not recruited and not overloaded will not get stronger. This type of training is referred to as High Intensity Training (H.I.T.). Additional information on H.I.T. training can be found at

<u>www.cyberpump.com</u>. Strength training goals will vary for each person; therefore, you must develop a program tailored to your goals:

Energy systems

Energy requirements are important considerations for an effective training and conditioning program. Once you determine these requirements you can factor in the amount of recovery time best suited to you.

The body uses three basic types of energy systems, and each functions differently. Your pre-set goals will determine which energy system you incorporate when designing the program. The energy needed for muscle contraction is derived from the breakdown of food into glucose and fatty acids to produce the chemical compound adenosine triphosphate (ATP).

- Phosphagen system is the metabolism of stored ATP and creatine phosphate. This system is used for high explosive activities with energy stores lasting less than 20 seconds.

- Anaerobic glycolysis is the metabolism of glucose and used for high intensity, moderate duration from 1-3 minutes.

- Aerobic glycolysis is the metabolism of glucose and fatty acids for activities lasting more than 5 minutes.

Rest periods

Rest periods influence the ability to generate greater muscle. The influence of intra-session rest intervals may have a profound effect on strength gains subsequent to short-term high-intensity training. To determine your necessary rest periods, the factors to be considered are volume of weight, intensity, and fitness level. To keep things simple: if you are a beginner, keep your rest period between exercises less than 30 seconds, more advanced person, between 60 and 90 seconds.

Overload

In order for a muscle to increase in strength, the workload it is subjected to during exercise must be increased beyond what it normally experiences (Darden, 1990). A person's rate of adaptation to training is limited and cannot be forced beyond the body's capacity for development. Individuals respond differently to the same training stress, so what might be excessive training for one person might be below the capacity for another (Darden, 1990). Muscles adapt to increased workloads by becoming larger, stronger, and by developing greater endurance. However, it is important to increase the overload gradually.

When do you increase your overload (or amount of resistance)? Only increase the overload upon the successful completion of 12 repetitions. At this point, the resistance can be increased by 2.5-5 percent, adding small but frequent increments that progressively stress the muscular system. By increasing the load in small increments, the chance for injury and over-training is lessened. A general guideline for reps:

Fitness level	Sets	Reps
Beginner	1	12-15
Intermediate	1	MMF
Advanced	1-3	MMF

Variety

Each of your training sessions should be challenging, but at the same time true to your intended goals and needs. When the body adapts to a stimulus, it works more efficiently and requires less energy. Each session should demand more from your body than the last session.

There are many ways to make sessions challenging and fun. For example, **changing the order of exercises**, mixing in different types of exercises, using a variety of equipment (machines, free weights, *etc.*), and varying the recovery intervals between sets are all effective methods for keeping it challenging.

Another example is to **increase the length of time to complete a repetition**. Raising the weight for 4 seconds and lowering for 8 seconds can be extremely challenging. As you become more adept at various training techniques, introduce advanced techniques. Examples of advanced training techniques are super-slows, breakdowns, and negatives. Such advanced techniques require more recovery between sessions; therefore, it is imperative to keep accurate records of your training.

Advanced Training Techniques
Super-slow training

Super-slow training requires lifting and lowering the weight in a deliberately slow motion using proper form, thereby eliminating all momentum. This training method is both physically and mentally demanding. Given this, it is recommended that you reduce the amount of weight you normally use for the given exercise by 25-40 percent. Once the weight has been modified, raise the weight for 10 seconds and lower it for 10 seconds, then raise the weight for 3-4 seconds and lower for 8-10 seconds.

Breakdown training

This technique is applied at the point of MMF by reducing the resistance to permit additional 3-8 repetitions. The amount of weight that is reduced varies, and is dependent on your body's comfort and the recovery between sets.

Breakdown Training Techniques

LENGTH OF REST INTERVAL	APPOXIMATE REDUCTION IN WEIGHT
5-10 seconds	30-40 percent
10-20 seconds	20-30 percent
20-30 seconds	10-20 percent
40-60 seconds	5-10 percent

Keep an accurate training log to help you maintain and gradually increase your progression. The training log should list the name of each exercise, the rest-time intervals between each set, the number and length of reps performed, and the date, time of day, and length of each rep. This is additional work, but the benefits will be seen on paper. Your results will be significantly greater with accurate record keeping.

Maintaining interest

To retain interest and enthusiasm for a training program, be creative while planning each of your training sessions. Change the order of exercises, number of repetitions and sets, training environment, and recovery period.

Do something new, or use different training equipment for each workout. Go outside for power walks, use trees (branches for chin-ups), bushes (as hurdles), and park benches (great for push-ups and dips). **The outdoors is a wonderful alternative to the weight room.** Rocks can also be great alternatives to dumbbells. If you are not a member of a gym, purchase a wide array of equipment such as exercise bars, bands, balls, and flexibility programs to use at home. Get involved in different aerobic and group classes such as kickboxing, sports conditioning, and dance.

Weight-bearing exercise routines can get old quickly; there is no rule that says a workout needs to be done with weights. Mix it up with aerobic exercises. Satisfaction with your program can be achieved by constantly presenting yourself with new exercises, challenges, and fun activities.

STRENGTH TRAINING FOR GENERAL FITNESS

A well-designed strength and conditioning program addresses all major muscle groups. Such a program produces overall strength development and can reduce the risk of muscular imbalance injuries. In each strength training session, the following exercises are recommended:

Exercise	Muscle group
Leg press	Total leg
Leg extensions	Quadriceps
Leg curls	Hamstrings
Pullover	Latissimus dorsi
Bench press	Pectoralis Major
Compound/seated row	Trapezius
Overhead/shoulder press	Deltoids
Triceps extension	Triceps
Barbell curl	Biceps
Abdominal crunches	Rectus abdominis
Lower back extension	Erector spinae

Educate yourself on all possible ways to perform these basic exercises. You will benefit greatly from such variety. The more ways that you add variety to you program will keep you motivated and making gains.

On days that you do not perform strength training, participate in some form of aerobics, flexibility exercises, play, or simply take time and rest. Create a well-designed, five-week program for general fitness such as the one below.

Mon.	Tue.	Wed.	Thu.	Fri.	Sat.	Sun.
Off	Full	Off	Off	Full	Off	Off
Full	Off	Full	Off	Full	Off	Off
Upper	Lower	Off	Upper	Lower	Off	Off
Off	Off	Full	Off	Off	Off	Off
Off	Full	Off	Off	Off	Full	Off

- Full body—All major muscles are trained (chest, traps, shoulders, lats, arms, torso, and legs); 12-15 reps.
- Lower body—All major lower body muscles are trained (quads, hamstrings, hips, and calves); 15-20 reps.
- Upper body—All major upper body muscles are trained (traps, shoulders, chest, lats, and arms); 12–15 reps.

A sample five-week workout is on the next page. Always warm up with light aerobic activity for a minimum of 5 minutes before beginning your workout.

**Week 1
(Full Body)**

Seated rows

Chest flys

Chest presses

Seated pull downs

Lateral raises

Shoulder presses

Biceps curl

Triceps extensions

Leg curls

Leg extensions

Leg presses

Ab crunches

**Week 2
(Full Body)**

Leg extensions

Leg adductions

Leg curls

Leg presses

Shoulder presses

Pullovers

Chin ups

Chest presses

Shrugs

Upright rows

Dips

Biceps curls

Week 3
(Upper)

Chest flys

Chest presses

Shrugs

Seated rows

Front raises

Shoulder press

Lat pull downs

Triceps extensions

Biceps curls

(Lower)

Leg extensions

Leg presses

Leg curls

Leg presses

Ab crunches

Lower back extensions

Week 4

Leg abductions

Leg curls

Leg extensions

Leg presses

Pushups

Upright rows

Shoulder presses

Lat pull-downs

Dips

Week 5

Lateral raises

Shoulder presses

Chin-ups

Chest flys

Dips

Shrugs

Triceps extensions

Biceps curls

Leg press (2 sets)

Training tips for the beginner

Your goal should be to perform all the reps with perfect form (slow, controlled movement with no bouncing) for the target rep goal. If you can do a few more reps, do them. Add 5% more weight the next time you do the same exercise. Each workout, try a few exercises with different equipment. This will make the workouts fun, and keep the muscles stimulated.

Training tips for the more advanced

Your goal is to maintain perfect form with every exercise and rep. Never sacrifice quantity for quality. Perform each set to complete concentric failure, and then lower the weight as slowly as you are able. The longer the muscle is under tension, more fibers are recruited and the stronger the muscle can become. If you can do more reps, do them. **The muscles will maximize their potential by going to momentary muscular failure.** During your next workout, add more weight, slow your rep count, change the order of exercises, change the rep range, or try an advanced technique as mentioned earlier.

Strength training

Your strength program should start each training session with a different muscle group; this will help promote muscle balance. Whenever a muscle is disproportionately stronger than its antagonist (the muscle on the opposite side of the joint), the latter is predisposed to injury (Westcott, 1996). Also, working a different muscle group each time reduces boredom and training plateaus that most people experience when strength training.

For example, if you started with a chest exercise on Monday, begin with a back exercise the next day you strength train. This is a great procedure for keeping the body in balance. The variations of exercises can be endless. For each session, alternate exercises between free weights, machines, cables, dumbbells, bands, or any other approach that offers variety. Another way to reduce monotony is to use circuit training and interval techniques. Circuit training is a series of exercises, performed in sequence, using several pieces of equipment. Circuit training provides a workout for all major muscle groups and has the potential of improving cardio-respiratory endurance, muscular endurance, and muscular strength (Katz, 1992).

Circuit training

Circuit training can strengthen the cardio-respiratory system by alternating lower body with upper body exercises. While forcing the heart to work harder, blood flow shunts up and down the body to the working muscles. Because of the short rest periods and the use of many muscles, the exercise intensity of circuit training is typically lower than that of regular training (Stone, 1987).

When designing the circuit, consider the specific parts of the body and components of fitness. The intensity should range from 40-0% of maximum strength capacity. The recovery-to-work ratio can be one minute of exercise, followed by 30 seconds for recovery. There are several variables to consider before starting:

Number of stations: Circuits to develop strength and cardiovascular fitness may have as many as 6-12 stations.
Duration of each station: For general fitness, each set should last 60-90 seconds per station.
Repetitions and recovery periods: Perform 12-20 repetitions using proper form.
Arrangement of stations: Either use the push/pull approach, or alternate between upper and lower body.

Interval training

Interval training divides endurance activity into hard and easy sessions. Aerobic interval training can be used to give a boost and to overcome plateaus. Interval training should be used when a high level of aerobic fitness has already been achieved. An example of an interval training plan might be to run 1/4 mile hard, and then run the next 200 yards at an easier pace. This sequence may be repeated 4-10 times. Interval training may contribute to faster running times and also increases the heart's stroke volume and pumping capacity.

Cross training

Cross training provides the added benefit of improved aerobic capacity. Examples of cross training might be bike riding 10 minutes, stair stepping, and easy jogging done consecutively. By using different muscles in different activities, the general conditioning effect of the exercise is increased and overuse injuries are decreased.

Aside from the potential for overuse injuries performing the same exercise for an extended period of time (30-40 minutes) can become boring. Cross training offers greater freedom and reduces the boredom factor. There is also evidence that individuals who cross train experience improvement in both short-and long-term endurance.

Integrating Aerobics with a Strength Training Program

When should aerobics be done: before or after strength training? For overall fitness, the order is a matter of preference. Forty-three subjects participated in a study to determine whether it was more beneficial to perform aerobics before or after strength training. Twenty-two subjects performed aerobics first, followed by strength training; 21 subjects performed strength training first, followed by aerobics. The results revealed that the order of the activities yielded little difference in benefits (Westcott, 1996). To apply these results, it is important to consider your training goal: endurance or strength and power? If endurance is the goal, aerobics should be done first, followed by strength training. When power is the goal, strength training should be done first, followed by aerobics.

Another pertinent question is, "Should I do aerobics the same day as strength training or on alternate days?" Collins' (1993) study concluded that adaptations to a combination of short-term endurance and strength training appear to be independent of whether endurance training is completed prior to or following a strength training session.

His study was conducted over a 7-week period. Twenty-three females and 11 males performed 10 strength training exercises for 2 sets of 3-12 reps. This was followed by endurance training consisting of 20-25 minutes of running at 60-90% of their maximum predicted heart rate. The participants were split into three groups: one group did endurance training prior to the strength training, the second group did the strength training first, and the third group was a control group. The test measured VO_2 max, bench press, shoulder press, arm curl, and leg press.

The results showed that the participants in the group that performed endurance training first had greater improvement in all five tests compared to the second group. However, the first group's relative improvement over the second group was marginal, showing that strength measures appear to be independent of whether endurance training occurs prior to or following strength training.

STRENGTH TRAINING FOR ADOLESCENTS

These days many young children are involved in numerous sports and activities. Strength training, flexibility, and cardiovascular programs can be a great asset to their health and life development. Studies have shown that many **children who engage in "fun" fitness programs at an early age tend to stay involved throughout their lives** (Miller, 1983). A child's involvement in an exercise program can also enhance self-esteem and promote positive attitudes. A study to measure the attitude levels of children pre-and post-involvement in a 7-week conditioning program—which included strength, endurance, and flexibility exercises— demonstrated that children's attitudes towards physical education, physical fitness, and lifelong exercise improved significantly (Wescott, 1992).

By participating in physical activities at an early age, kids learn kinesthetic skills, *i.e.*, knowing how and where their bodies move and react to different physical stimuli. Fitness activities can also improve self-confidence and help develop such skills as leadership, commitment, sacrifice, and. When children have the physical ability to do certain tasks (run a mile, or do sit-ups), they develop respect and appreciation for their bodies. More importantly, an early introduction to health and fitness tends be a trend continued for life.

The first ingredient for kids' fitness is fun. We already know that children will perform all kinds of work, providing they have fun doing it. Therefore, be creative when developing a fitness program for kids, just as you would for adults.

There's no need to rush a child into any type of fitness program; a gradual progression is the best approach if goals are to be met. The slow approach helps to avoiding injury and feelings of failure in the child. Constant monitoring, education, and assessment (fitness tests, evaluations, and observation of tasks) are critical continuation of a healthy, active lifestyle.

To craft a fitness program for a child, obtain a complete medical and physical exam, plus parental consent. Always start each session with a warm-up period, and end each with a cool-down period and a series of flexibility exercises.

I recommend you take and pass children's CPR and first aid class as well as research how best to train children.

Adolescent obesity is an epidemic today. Huge food portions, fast foods, and sedentary hours parked in front of the television or computer are all contributing to this epidemic. Children and adolescents need to get moving, get outdoors, and off the couch.

Training adolescents is an excellent opportunity to turn this epidemic around. Strength training for adolescents was once considered to be unproductive and posed a high health risk. Today, however, strength training has been found to be extremely beneficial—with proper guidelines and instruction.

Even if adolescents are not exposed to health and fitness at an early age, they can always benefit by beginning a fitness program. Adolescents who begin or continue to participate in recreational activities and strength training can improve gross and fine motor skills, balance, and coordination. In addition, adolescents who are active are less prone to becoming couch potatoes. Active adolescents also develop better self-esteem, social skills, and life fulfillment.

However, there may be risks involved when training adolescents. The most common risk is the possibility of damaging the epiphyseal plates. Children and adolescents are still growing. In children, the area of bone surrounding the growth plate has not yet achieved its mature strength and is predisposed to injury. With a safe and properly designed program, the potential for injury is reduced. Another possible cause of injury is lifting weights that are too heavy. The ACSM recommends 12-15 repetitions per set; one-repetition maximums should never be attempted. Over-training muscles inadequate proper recovery time can also lead to injury.

Accidents are another area of concern. Strength training in a safe environment and implementing a program with proper progression can lower these risks. Adolescents who are not taught proper lifting techniques are at greater risk for injury, such as having a weight fall on them or straining a muscle. Keep the training environment free of clutter, and make sure the adolescent or child knows and understands the rules, and knows what is expected. Most importantly, make the process enjoyable.

Training frequency should be 2-3 times per week, with each session lasting no more than 45 minutes. Start the session with no resistance, and then slowly progress to resistance without maximal lifts. Perform one set of 12-15 repetitions incorporating all major muscle groups. Only when the adolescent can perform 15 reps with proper form you should increase weight 2-5%.

STRENGTH TRAINING FOR SENIORS

The body's gradual biological functional declines are the result of aging. Aging affects all of us, but a rapid decline need not. Related declines due to aging are decreased cardiorespiratory function, increased body fat, and musculoskeletal fragility. That's the bad news.

The good news is that today more individuals over the age of 60 are seeking to improve their health and quality of life through some form of resistance exercise (Thompson, 1994). **Research indicates that a well-designed, progressive weight training program will increase overall fitness and provide greater functional mobility as one ages.** A complete medical evaluation and physical assessment should be mandatory prior to commencing the program. Seniors must seek advice from a physician and not begin vigorous physical activities that could introduce orthopedic problems.

Once an exercise program prescription has been approved, apply proper progression and emphasize proper form. Start with compound movements (movements that incorporate multiple muscle groups) to develop strength and balance, and perform one set of 12-15 repetitions.

As women age, they may gradually lose 20-30% of their total bone mass. Weight training can slow or even prevent this process, as physical activity promotes increased bone mineral density. Exercise is a key strategy for preventing and treating osteoporosis (Katz, 1998). In addition to increased bone density, seniors who strength train are likely to increase muscle strength and metabolism, reduce body fat, lower blood pressure, and improve lower back strength. A study by Westcott (1996) demonstrated an increase in muscle strength by an average of 40% in seniors who engaged in an 8-week program.

Arthritis is an equally important consideration. Arthritis, which refers to different diseases, is the top-cause of disability in the United States among seniors, and osteoarthritis (a degenerative joint disease) is the most common kind of arthritis (Dinubile, 1997). Inflammation of one or more joints is the main characteristic of arthritis. As a result of inflammation, pain and stiffness may be present in adjacent parts of the body, such as muscles near the joints, particularly weight bearing joints. Strength training increases mobility of the joints and produces greater strength of the muscles that support and protect the joint. Regular exercise can significantly decrease joint swelling caused by arthritis.

Research indicates that the aging musculoskeletal system retains and improves its responsiveness when exposed to a regular, progressive resistance regimen (Thompson, 1994). Vary each training session with flexibility exercises and make the time enjoyable. Train all the major muscles of the body to develop strength and balance. Lastly, get out and move. **Surround yourself with youth.** Volunteer at schools. By having a happier mind set and a healthy body, you life will be dramatically enhanced.

STRENGTH TRAINING AND PREGNANCY

The American College of Obstetrics and Gynecology stated in 1997 that moderate exercise helps maintain cardio respiratory and muscular fitness throughout pregnancy and the postpartum period. Women who are engaged in a fitness program, and who stay physically active during and their pregnancy often experience an easier pregnancy. Studies have shown that when compared to sedentary pregnant women, women who exercised underwent shorter labor, required less medical interventions during delivery and post-pregnancy, and recovered more rapidly.

Many believe that physical exercise may contribute to the growth of the placenta, increasing and making more nutrients available to the baby. However, there are numerous physiological and anatomical adaptations caused by the pregnancy that affect a woman's ability to exercise. A pregnant woman's body will respond differently to exercise. Changes in coordination and balance, body position during exercise, increased nutritional requirements, and the potential for back and pelvic pain all must be considered (Arauj, 1997).

Prior to undertaking an exercise program, consult with an obstetrician when determining the course of the program. Women who have exercised prior to pregnancy can still continue their fitness programs, but with a decrease in intensity. This decrease is recommended since the women will have a decreased amount of oxygen available to them, as they now share oxygen and nutrients with the fetus. The decrease will have a negative effect on the woman's performance, but not on overall health benefits received. The exercises must be self-limiting and individually paced. In other words, pregnant women need to act in a sensible manner when training and should not attempt to over-exert the body.

In the early stages of exercise, a pregnant woman may experience a decrease in performance due to nausea, fatigue, vomiting, and other physiological body-shape changes. Hydration must be emphasized. Always remain hydrated by drinking water every 10-15 minutes before, during, and after exercise.

Regular exercise at a mild to moderate intensity level (65-75% of maximum predicted heart rate MPHR) performed 3 times per week is recommended, and should be continued throughout pregnancy. Non-weight bearing exercise can be maintained at a moderate to high level of intensity.

The intensity of exercise should be based on the perceived exertion scale of 0-10 (where 0 is the easiest and 10 the most difficult). A pregnant woman should not exercise to exhaustion, so exercise at level 5-6 is recommended. She should be able to maintain a conversation while exercising and not gasp for breath. If the ability to continue a conversation is difficult, decrease the intensity level, and maintain a comfortable pace for about 20-40 minutes.

Contra-indications when exercising are important to understand. Avoid the supine position after the first trimester as it may compress the uterine vein. If contractions occur, reduce activity. If contractions persist, stop exercising and consult a doctor.

Step aerobics and slide training should be avoided due to loss of balance and joint laxity (ACOG, 1997). A potentially dangerous situation during weight training exercises for everyone is the tendency to hold one's breath. Doing so increases the intrathoracic pressure and is referred to as the valsalva effect. The increased pressure on the chest may be enough to slow or inhibit blood flow from the veins to the heart, possibly resulting in increased blood pressure. This, of course, is particularly detrimental to a pregnant woman. Always encourage and promote regularly breathing when lifting weights.

Key points for exercising while pregnant:

- Train for muscular endurance.

- Target stretching quads, hamstrings, calves, and pectorals.

- Reduce stretching during the last trimester.

- Avoid the supine position after the first trimester.

- Perform 1-2 sets of exercises with reps ranging from 15-20.

Machines and equipment to be avoided

AEROBIC EQUIPMENT	PROBLEM	ALTERNATIVE
1. Recumbent bike	Sciatica or leg pain	Walk
2. NordicTrak	Lack of balance	Rower

Machines and equipment to be avoided (Continued)

3. Stairmasters	Causes back ache	Walk
4. Treadmills	Lack of balance	Use handrails

NAUTILUS EQUIPMENT	PROBLEM	ALTERNATIVE
1.Chest fly (lying)	Body in supine position	Use incline press
2. Super pullover	May cause a dislocation of shoulder	Pull-downs
3. Compound row	Places compression on abs	Free weights
4. Low back	May cause back pain	Stretch low back
5. Abdominal	Places compression on abs	Pelvic tilts
6. Prone leg curl	Causes abdominal pressure	Seated leg curls

Hammer Strength	Problem	Alternative
1. Low row	Places compression on abs	Free weights
2.Hip and back	Places body in supine position	Pelvic tilts
3.Crunches	Puts client in supine position	Pelvic tilts
4.Rear deltoids	Puts client in prone position	Rowing back

Free weight	Problem	Alternative
1.Supine leg press	Places body in supine position	Other leg press
2.Supine/decline press	Places body in supine position	Hammer
3.T-bar	Places body in prone position	Rowing back

Other exercise-induced physiological changes that affect a pregnant woman are resting heart rate increases and maximal heart rate decreases. As a result, target heart rate formulas cannot be used for pregnant women.

Pregnant woman experience decreased oxygen available for aerobic activity, increased joint laxity, a shift in center of gravity, and poor balance control. They should avoid pivoting turns and jarring movements and use caution when getting on and off treadmills.

Also, pregnancy places added caloric demand on the body (ACOG, 1997). Proper carbohydrate intake and hydration is required. Small balanced meals (lean protein and carbohydrate) should be eaten 1-2 hours before exercise to avoid hypoglycemia.

Musculoskeletal Complaints

Complaint	Solution	Avoid
Leg cramps	Calf stretches	Deep massage
Swelling	Elevate legs	Aerobics
Dehydration	Stretch	Exercise
Back pain	Stretch	Excessive bed
Gluteal pain	Stretch	Prolonged walking
Sciatic pain	Hamstring stretch	Exercise

PREVENTING INJURIES AND TREATMENT

There is always a potential for injury with any type of physical activity. Training in a safe environment with controlled movements, and slow progression that is consistent with abilities and limitations will reduce or eliminate the occurrence of an injury.

Injury prevention is your foremost priority. Participating in pre-season and in-season muscle conditioning and strength training programs reduce injuries. In addition, periodization (applying the principles of overload and progression gradually, with adequate recovery periods) can also reduce potential over-training and injuries by varying the volume, intensity, and selection of exercises during a training program.

Understanding signs of distress will enable you to respond immediately to a given situation. Some signs of potential injury are redness or suddenly flushed skin, swelling, over-heating, the loss of physical function, and, of course, complaints about pain. The time frame to care for an injury is from the time of injury up to 72 hours thereafter. If you or your client sustains an injury, follow these important injury-care steps:

- **Protection**—To prevent greater harm to the injured area, use a form of a cast, crutches, brace

- **Rest**—First and foremost, modify activity level

- **Absolute rest**—Complete rest from current activities

- **Related rest**—Pain-free activity, reduce activity in sport, alternative training activity

- **Ice**—Reduces tissue damage by decreasing swelling, spasms, and nerve conduction

- **Compression**—Helps control or reduce the amount of swelling

- **Elevation**—If possible, keep injury site above the heart. This will limit fluid pooling

Ice application:

1. Immediately apply ice pack to injury; place thin towel between skin and ice.

2. Hold or securely fasten ice pack to injury site.

3. Elevate the site above heart.

4. Apply pack for 20 minutes, and then remove.

5. Reapply ice pack every 2 hours, and continue for 24-72 hours depending on the swelling.

Common sport-related injuries

Following this manual's protocol for strength training and exhibiting common sense while exercising can reduce injury. Nonetheless, injuries can and do occur. Some common exercise-induced injuries include:

Shin splints: Small tears in the front lower leg muscles where they are attached to the tibia. Shin splints can occur anteriorly (affecting the tibialis anterior) and posteriorly (affecting the soleus). This injury is a result of muscular imbalance between the calf and anterior tibialis muscles.

Prevention: Run on soft surfaces, wear well-padded, shock-absorbent shoes, and slowly advance into exercise. Also, stretching the muscles in the front and the back of the lower leg may help in prevention of the problem.

Runner's knee (chondromalacia patella): An inflammation manifested in the cartilage under or around the kneecap. This can result from weak quadriceps, from direct trauma to the knee, or from overuse.

Prevention: Prevent stress on the knee by avoiding excessive amounts of running, jumping, aerobic dance, and stair climbing. If the injury persists, anti-inflammatory drugs may be needed to alleviate pain.

Achilles tendonitis: An inflammation of the connective tissue between the muscle and bone usually occurs 1-3 inches above the heel.

Prevention: If you or your client complains of tendon pain, stop exercising immediately; this will at least prevent further irritation and reduce the potential severity of the tendon damage. Achilles tendonitis can be managed with protection, rest, ice, compression, and elevation.

Over-training

Over-training is a psychophysiological malfunction: the demonstrated inability of the athlete to adjust to the demands of training stress. This is just one of several indicators that shows an athlete is not responding positively to the current training. Over-training is usually a result of a failure to understand the relationship between frequency, intensity, and the duration of the training program. Over-training may be marked by a plateau or drop in performance over a period of time (Fleck, 1982).

The best markers are the most obvious: decreased performance, low energy levels, perception of fatigue, and the inability to improve from training (Fahey, 1997). You must carefully balance training intensity and volume with proper rest between training sessions. A methodical plan can greatly reduce the risk of over-training. Often, as the muscle develops and the intensity is high, you may experience soreness and stiffness, but will continue to train. You must rest, so repeated soreness, which will slow your gains and possibly sour you on future exercise. **The keys to maximum gains are quality (not quantity) of exercise and adequate rest.**

Be aware of individual responses to the same training stimulus. By repeating the same training, the body adapts to the stimulus by becoming more efficient. This adaptation response requires constant alteration/tweaking of training routines, drills, and training environment.

If you are a trainer, inquire about your client's personal and work lives. An understanding of what they deal with on a day—to—day basis will help you design an ideal program. If the client's job requires travel and long hours, stress levels and fatigue tend to be high. Common causes of over-training include performing extra workouts to make gains, not allowing sufficient time to recover, and starting a training or exercise program with over-zealous enthusiasm.

There may be other stressors in the your client's life that can affect performance:

For example, you may be under a deadline at work, may need to replace the roof on your house, or perhaps your young teen just started dating. These problems could cause you to get only four hours of sleep the previous night, or turn to junk food. All of these will hinder performance. Remember: life's stressors affect everyone differently. What may panic one person doesn't necessarily hit the radar screen of another.

It is amazing how just a few stressors can and will affect performance. Work with your client to find ways to ease the pressure. The mind and body both need to be strong and healthy for success.

Symptoms of over-training (Fleck, 1982)

- Physiological
  - Muscle soreness and stiffness
  - Decrease in performance, body weight
  - Irregular sleeping patterns
  - Loss of appetite
  - Elevation of resting and/or post-exercise heart rate
  - Elevation of blood pressure
  - Increase in body fat percentage

- Psychological:
  - Depression
  - Inability to relax
  - Loss of motivation

Frequently monitor for these symptoms of over-training. If they are evident, immediately stop activity and re-evaluate the program.

The best way to deal with over-training is to immediately decrease the intensity of training. As the intensity is reduced, the work volume can be increased. Slowly a cyclical of sessions: easy/moderate/hard, easy/moderate/hard, etc. One or two days of intense training should be followed by an equal number of easy days.

Other suggestions to help get back on track include:

icing after workouts, or ice baths, massage, vacation and mental relaxation activities.

THE ASPIRING PERSONAL TRAINER

Your goal is learning and teaching others how to make effective lifestyle, health, and fitness changes. You most likely have a great job—or at least one that's satisfying and pays the bills—and have no reason to alter that aspect of your life.

Other readers may have realized that the health and fitness industry may need you, as I did. You may find that you have a unique understanding of fitness and health. You may feel drawn to continue your fitness and health education. You may feel a need to share with others what you have learned. You may have the urge to carve a new career.

If you are experiencing these urges, this is your section! It will present the general guidelines of what you should do to pursue a highly sought-out career as a personal trainer. For those readers who are not interested in this career path but intend to train with a coach or personal trainer, it will offer a better understanding of how a personal trainer should assist you.

Like many others, I was drawn to the health and fitness industry purely by happenstance; it wasn't a conscious career decision. My love of health and fitness grew from pursuing changes in my own life.

Unbelievably, my came to me while mowing the lawn! I instantly knew that I wanted to share my passion for fitness and exercise. In order to effectively impart my knowledge, share my fitness passion with others, and be recognized as a leader in the fitness and health industry, I returned to college to earn a degree in Sports Management and Physical Education. Additionally, I earned my Certified Conditioning Specialists certification with the National Strength Professionals Association, one of the top certification organizations in the country. I then immersed myself in the industry. I interned at health-related facilities, bartered with clients as a personal trainer, and gave free lectures and workshops—all with the sole intention of learning as much as possible. Today, even though my career path is chosen and well established, I continue to stay current with the ever-changing research in my field, and always keep an open mind to new ideas and research.

For readers seeking a career as a personal trainer, your success will depend upon learning effective tools about fitness, mastering people skills, and—most importantly—retaining clients. These skills include an understanding of the following key elements:

- Successful business strategies for the personal trainer
- An overview of anatomy and physiology
- Fitness testing
- Physical assessments
- Strength training principles
- How to design and implementation of an overall strength and conditioning program

Successful business strategies

To succeed as a personal trainer, you need excellent personal relations skills. This is all about customer service. Successful personal trainers must have excellent social and business skills: the ability to work with all types of people and personalities; the ability to teach, promote, and inspire all types of people; and the ability to convey information clearly and concisely. Acquiring and honing these skills is an ongoing educational process. Personal training is far more involved than merely counting reps; it is about getting results for your client, and keeping your client satisfied. If you have a sincere and insatiable desire to keep learning about health and fitness, *and* enjoy dealing with people, then you will succeed as a personal trainer.

One contributing factor to my success as a personal trainer is that I genuinely enjoy life. If personal training is your intended occupation, then believing in yourself, being passionate about your job, being genuinely interested in others, and waking each day wanting to make a difference in someone's life is the path to success. Few people can honestly say that they have had a direct impact on someone's life. A career as a personal trainer will ensure that you do. If you want to make a difference in positively changing people's lives, while owning a business that allows you to make your own schedule, then personal training is the answer. An ancient Chinese proverb advises, "A man without a smiling face must not open a shop."

If you are planning to venture into personal training, it is essential to get professionally certified. Certification will provide you with the necessary knowledge of strength training and program design. It will also give you important information on preventing clients from injury. I highly recommend the National Strength Professionals Association (NSPA), one of the top certification organizations in the United States.

A college or university degree in physical education and/or health education will educate you on how the human body works. Afterwards, enroll in professional workshops and lectures related to the training industry to stay updated and fresh on new trends and ideas. Simply knowing how to do a barbell curl will not cut it as grounds for a personal trainer; it is very important to have academic training to back up your fundamental knowledge. Many exercise methods that were popular in the 1980s have been fickle, so don't fall behind the curve by spouting potentially dangerous exercise methods; it can hurt your business and your reputation. Not all fitness industry experts agree on the best methods to use. By familiarizing yourself with all fitness theories, your client will come to rely on you as an essential source of information. To bolster this argument, ascertain that your information is accurate and up-to-date; those who think they know everything are doomed to fail.

As a personal trainer, ask yourself the following questions: What distinguishes me from other personal trainers? Why should someone hire me? What are the characteristics of a successful personal trainer? You should be able to easily answer all of these questions. In my view, successful personal trainers should be knowledgeable, personable, pro-active, energetic, sincere, great listeners, and serve as role models. Do you have these attributes?

Another important consideration is your appearance. Although a great physique is not a prerequisite for a trainer, it is important to understand that you represent a person who is living a healthy lifestyle. Potential clients will naturally seek out those trainers who "look the part." Your attitude and appearance are a great motivator and inspiration for others to follow.

I have enumerated guidelines that will help you achieve your desired success as a personal trainer. As with most new ventures, the first thing you need is a feasible plan. Developing a viable business plan is essential.

The first step in developing such a business plan is to undertake a "S.W.O.T. Analysis" (a methodical marketing tool to examine your strengths, weaknesses, opportunities, and threats). Once completed, this analysis will determine the best-suited activities for the aspiring personal trainer. Performing the S.W.O.T. Analysis is fairly straightforward: honestly assess each aspect. For example, my *strengths* are strong people skills and public speaking, but my *weakness* is poor organization. Obesity in both adults and kids provides me the *opportunity* to train, and the *threat* is too much demand for my services. Explore ways to improve and strengthen each category and item. Your business plan will work best when it is reviewed and revised at least twice a year. If need be, seek counsel from a professional marketing or businessperson.

Once a plan is determined, the next step is finding clients, oftentimes the most daunting hurdle for a novice businessperson. (In reality, procuring your first three clients can be quite easy; you'll see why below.

The most important professional help you will need in order to get your new business off of the ground is the assistance of a lawyer, an accountant, and a banker. All three of these individuals are essential to the successful operation of a business. A lawyer is needed to help set up your company (sole partnership, partnership, or subchapter S corporation) and to write and file all legal documents (waivers, contracts). An accountant will handle bookkeeping, tax issues, and financial planning matters. Finally, a banker is needed to prepare loans, provide financial information, and set up new accounts. As a bonus, if you explain to each of these professionals that you are forming a personal training enterprise, you may be able to offer your training services in exchange for their services! There are your first three clients! Now you are ready to set up your office.

Setting Up a Home Business

Before setting up your home office, check with your local town and county authorities to ascertain that there are no regulations preventing you from operating as a business from your home be legal to display a business sign in front of your home. If you are working from your home, bear in mind that first impressions are lasting impressions, especially if you intend to bring clients into your home. There are several ways to ensure that your separate your home life from your business life.

Install a separate business phone line and an answering machine or service. This will inform the public that you are a professional.

Invest in general office equipment. Purchase a computer with basic business software, a copier, and a fax machine. These tools will help you manage your business, both yours and your clients' finances, and you can even save money by designing and printing your own business cards and business forms.

Establish liability insurance. You are an independent contractor; you deal directly with people in situations where injury is a possibility and liability insurance is mandatory. (A good source for trainers' insurance is IDEA, at 1-800-999-IDEA. For approximately $200 a year, you can get $1 million in coverage.)

"Dress for success." If you want clients to seek out your services and pay you the big bucks, then this phrase must to be adopted. Under most circumstances, you only get one opportunity to make an impact, so make it a positive one. Creating a professional image is critical for business success. We've all seen trainers who are fat, physically inept, dressed in tank tops, sporting flip-flops, and wearing their hats on backwards! These are not examples of professional marketing. Here's a better professional image: men should be clean-shaven; women should wear minimal make up and jewelry. Every personal trainer should wear wrinkle-free clothes, clean shoes, and deodorant. Professional marketing requires your company name and logo printed or embroidered on shirts and jackets or other appropriate attire. Why would you advertise Nike or Adidas instead of promoting your new brand? Your logo will serve as free and continuous advertisement, and the marketing expenses can be used as a tax deduction.

Clearly, not all male trainers should be Mr. Olympia wannabes, nor should female trainers be stick-figure models. Yet, it is always better to be physically fit and trim; a client will be motivated when the trainer looks fit and lives a healthy lifestyle. Your aim is to educate and inspire people. If you don't look the part, you are facing an uphill battle.

Have a cell phone. While a cell phone is essential for obvious reasons, be cautious as to when and how you use it. It is not uncommon to spot trainers "training" their clients while chatting on their cell phone. If I were paying $50 an hour, I'd expect to be paying for my trainer's undivided attention. Clients hire trainers for a set block of time; show them respect and give them that undivided attention. Improper or inconsiderate cell-phone use (talking on the phone during a session) projects an unprofessional image and demonstrates that you are not invested in their training success.

Other considerations. Do you plan to travel to people's homes for training sessions? Do you plan to use your home as a base to train clients, the gym, or both? If you plan to travel, you will also need to purchase some basic equipment to take to clients' homes, such as exercise balls, floor mats dumbbells, and body bars. All are fairly lightweight and easily transported. This way, you can tell home clients all they need is a space to workout.

To maintain clients and to keep you focused, variety in training is essential. Every personal trainer knows that there is no single way to make muscles contract and flex, or to increase the heart rate. Developing muscles and getting lean can be achieved by using a variety of equipment in varying manners. For example, use a staircase; walking up and down stairs can serve as a great warm-up. Likewise, resistance training can be performed in many ways. For instance, curls can be done with laundry detergent bottles, or overhead presses can be done with bags of mulch. As long as you are well versed in anatomy, its functions, and range of motion, there's no reason why you can't be an innovator.

Seeking advice from people with great physiques can sometimes be misleading. Oftentimes, those who are self-absorbed and preoccupied with their physiques may not be the best representatives for the personal trainer industry. Often their "look" is derived from a combination of good genetics or muscle-enhancing steroids. These people tend not to care what their health will be like in their 60s, 70s, or 80s; they live for the moment, not the lifetime. Embarking on a healthy lifestyle is a lifetime commitment, and successful personal trainers are able to educate and motivate their clients to achieve this. It is human nature to want quick fixes. A healthy lifestyle, however, takes a lifetime and often benefits are not immediately visible. This is one of the core challenges for the personal trainer.

Personal training is, after all, about customer service. You are selling your personality and your knowledge. Successful personal trainers make lasting impressions by being pleasant, punctual, and attentive to their clients' needs. Let me repeat that last phrase again: **Attentive to their clients' needs.**

Allow me to list my commandments to succeed in life and in your training career:

- **Always give 100%.** Are you having a lousy day? Then pout in your car, not during your client's training session. You are being paid to motivate your client, not to sound off. No matter what, always give 100% of your effort. If you feel that you are unable to give your client 100%, give them a session gratis. **Remember that you are in the customer-service industry.**

- **Be honest.** Nothing ruins any relationship more than a lie. If you are asked a question and you don't know the answer, don't make up some preposterous answer. Instead, promise to determine the answer and get back to the client immediately. A simple, "You know, that's a great question, but unfortunately one that I don't know the answer to off the top of my head. Let me research it and get back to you," usually does the trick.

- **Become a great listener.** Listening is an essential skill in the personal training business. If you want to succeed, you must learn to actively listen. By attentively listening to your client, you acquire the essential information needed to develop the appropriate program for your client. Use active listening techniques: ask questions and truly listen to the answers; listen for expressed attitudes in your client's responses, such as his or her fears, hopes, and needs. Be attentive to the client's fitness interests, and try to fulfill them. Don't try to impress your clients with big industry words. Don't brag; your reputation should precede you. Focus on listening rather than expounding. It is imperative to create an easygoing, but strong, line of communication between your and your clients.

- **Become a great motivator.** If you cannot motivate people, your success as a trainer will be limited. Constant encouragement and re-enforcement should be offered for all improvements, no matter how small.

- **Avoid mediocrity.** If you set your standards low, your success will follow suit.

- **Refuse to procrastinate.** We have all been there before, and said it before: "I'll do it tomorrow." It is human nature to procrastinate occasionally, and there is nothing inherently wrong with doing this once in a while. However, if you are a chronic procrastinator, it will affect your business to the point of failure. Learn time management. Don't put off projects that seem difficult to resolve by opting for the easily completed task. You may find that your days are packed with appointments, hindering your ability to complete another task. Begin each day with a "to-do" list and rank the items in order of priority, not effort. Refer to this list often and keep on track. An opportunity tomorrow could be wasted by yesterday's work. If you have lousy time management skills, enroll in a course to help with this, or purchase a book on the subject; it is a rectifiable problem.

- **Get organized.** Less energy is required when you are organized.

- **Avoid burning bridges.** Here's one example of how to lose a future client: While driving, someone cuts you off in traffic; you respond by honking angrily and offer a few choice words and/or hand gestures. Is it so inconceivable that this person could be the same client you are about to meet for the first time? Improbable? Possibly. But you never know. Maintaining composure is an important aspect of your life and business. Even if you end up disliking one of your clients or have a run-in with some other professional acquaintance, you shouldn't burn bridges with these people. One day they may be of use to you, either as a contact or as a source of referrals

- **Make tardiness unacceptable and make yourself accessible.** A client is paying you for your time. It is unacceptable to be late. With punctuality, you demonstrate to your clients that you respect their time and money. Unfortunately, circumstances will occur which we cannot control. As discussed earlier, eliminate potentially aggravating your client by purchasing a cell phone. It will make you more accessible and allow you to maintain contact with your clients. If you will be more than a few minutes late, call and inform the client. This shows that you respect the person's time.

- **Recognize that you live in a fishbowl** Never make people question your character. Your best advertisement is a happy customer. People seeking the help of a personal trainer will ask a friend's advice before venturing through the Yellow Pages. Good word-of-mouth is worth its weight in gold. Again, let your reputation precede you.

Typically, personal training sessions last one hour. Make the time enjoyable, productive, and challenging for the client. Mixing up the sessions can enhance a client's motivation and can turn a short-term client into a long-term client. Your goal is to minimize the common exercise dropout rate among individuals who begin a vigorous exercise program. According to studies on exercise adherence (Carron, Hausenblas & Mack, 1996; Dishman, 1987; and Sallis & Hovell, 1990), the rate of dropouts and relapses in exercise programs is 50%. This is a startling percentage. As a personal trainer, recognize the significant potential of your client to drop out; therefore, you must keep the client focused on the bigger picture and the goal of a healthy and physically fit lifestyle.

A personal trainer is not unlike a psychologist. You must be able to explore, recognize, and pinpoint an area of distress; then apply your knowledge, and hopefully help to solve your client's problem. You need to assist your client in eliminating obstacles that may prevent continuation of the prescribed exercise program. It is your job to initiate "interventions" for first-time exercisers and those who have habitual relapses, by teaching them skills for coping with factors that become obstacles to their long-term fitness goal (Simkin & Gross, 1994). This can be accomplished in a number of ways. Every few weeks, sit down with your client and review your client's goals. Some clients may need to redraft their goals. If so, this is fine. Goals are not set in stone and can be modified to accommodate your client. Obviously, it is preferable to help the client achieve the goals, but not by causing the client distress. Most often, setting smaller, attainable goals is the best approach for keeping your clients motivated and focused on reaching the larger goal.

Most clients become disappointed because they are not getting the overnight results they expect. Others do not know how to push themselves, or lack the knowledge and motivation to effectively train themselves. However, those who hire a personal trainer are one step closer to admitting that they need some form of assistance with their goals. It is important for you to keep your clients accountable and responsible for their choices towards living a healthier life.

A regular review of goals and progress can keep the client motivated and focused on his or her goals. Leaving telephone messages or sending affirmations and tips by electronic or regular mail are great ways to keep a client on track as he or she deals with life's demands. The client will certainly appreciate the contact, and it will help to eliminate the feeling that the client has been given a plan and then left to battle through it alone.

You must also emphasize your own availability if a problem should crop up. Have flexible hours and encourage them to contact you in such instances. Of course, you don't want your clients calling you in the wee hours of the morning, but there should certainly be a sense of loyalty and commitment offered to them should they need to contact you outside of their paid session.

The key factor to your success as a personal trainer is the ability to develop both a personal and professional relationship with your client, and have a great time with the process!

APPENDIX A

Physical Fitness Profile Evaluation

Norms: Women 18 to 25

Rating	3-minute step test	Sit and reach	Sit–ups
Excellent	72-83	24-27	44-55
Good	88-97	21-23	37-41
Above average	100-106	20-21	33-36
Average	110-116	18-19	29-32
Below average	118-124	17-18	25-28
Poor	128-127	14-16	20-24
Very poor	142	8-13	4-17

Norms: Women 26-35

Rating	3-minute step test	Sit and reach	Sit–ups
Excellent	72-86	23-26	40-54
Good	91-97	20-22	33-37
Above average	103-110	19-20	29-32
Average	112-118	18	25-28
Below average	121-127	16-17	21-24
Poor	129-135	14-15	16-20
Very poor	141-154	8-13	1-12

Norms: Women 36 to 45

Rating	3-minute step test	Sit and reach	Sit–ups
Excellent	74-87	22-25	34-50
Good	93-101	19-21	27-30
Above average	104-109	17-19	24-26
Average	111-117	16-17	20-22
Below average	120-127	14-15	16-18
Poor	130-138	11-13	10-14
Very poor	143-152	6-10	1-6

Norms: Women 46-55

Ratings	3—minute step test	Sit and reach	Sit–ups
Excellent	76-93	21-24	28-42
Good	96-102	18-20	22-25
Above average	106-113	17-18	18-21
Average	117-120	15-16	14-17
Below average	121-126	14-15	10-13
Poor	127-133	11-13	6-9
Very poor	138-152	4-10	0-4

Norms: Women 56 to 65

Ratings	3-minute step test	Sit and reach	Sit–ups
Excellent	74-92	20-22	24-36
Good	97-103	18-19	18-22
Above average	106-111	16-17	14-16
Average	113-117	14-15	11-13
Below average	119-127	12-13	6-10

Norms: Women 56 to 65 (Continued)

Poor	129-136	9-11	2-4
Very poor	142-151	2-8	0-1

Norms: Women over 65

Rating	3—minute step test	Sit and reach	Sit–ups
Excellent	73-86	20-22	24-36
Good	93-100	18-19	18-22
Above average	104-114	16-17	14-16
Average	117-121	14-15	11-13
Below average	123-127	12-13	6-10
Poor	129-134	9-11	2-4
Very poor	135-151	2-8	0-1

Norms: Men 18 to 25

Ratings	3-minute step test	Sit and reach	Sit–ups
Excellent	70-78	20-26	50-60
Good	82-88	18-20	45-48
Above average	91-97	17-18	40-42
Average	101-104	15-16	36-38
Below average	107-114	13-14	32-34
Poor	118-126	10-12	26-30
Very poor	131-164	2-9	12-24

Norms: Men 26 to 35

Ratings	3—minute step test	Sit and reach	Sit–ups
Excellent	73-79	20-25	46-55
Good	83-88	18-19	41-45
Above average	91-97	16-17	36-38
Average	101-106	15-16	32-34
Below average	109-116	12-14	29-30
Poor	119-126	0-12	24-28
Very poor	130-164	2-9	6-21

Norms: Men 36 to 45

Rating	3—minute step test	Sit and reach	Sit–ups
Excellent	72-81	19-23	36-50
Good	86-94	16-17	29-33
Above average	98-102	14-15	25-28
Average	105-111	12-13	22-24
Below average	113-118	10-11	18-21

Norms: Men 36 to 45 (Continued)

Poor	120-128	7-9	13-17
Very poor	132-168	1-6	4-12

Norms: Men 46 to 55

Ratings	3—minute step test	Sit and reach	Sit–ups
Excellent	78-84	19-23	36-50
Good	89-96	16-17	29-33
Above average	99-103	14-15	25-28
Average	109-115	12-13	22-24
Below average	118-121	10-11	18-21
Poor	124-130	7-9	13-17
Very poor	135-158	1-6	4-12

Norms: Men 56 to 65

Ratings	3 minute step test	Sit and reach	Sit–ups
Excellent	72-82	17-21	32-42
Good	89-97	15-17	26-29
Above average	98-101	13	21-24
Average	105-111	11	17-20
Below average	113-118	9	13-16
Poor	122-128	5-7	9-12
Very poor	131-150	1-5	2-8

Norms: Men over 65

Ratings	3-minute step test	Sit and reach	Sit–ups
Excellent	72-86	17-20	29-40

Norms: Men over 65 (Continued)

Good	89-95	13-15	22-26
Above average	97-102	11-13	20-21
Average	104-113	9-11	16-18
Below average	114-119	8-9	12-14
Poor	123-128	5-7	8-10
Very poor	133-152	2-4	2-6

Modified from Y's Way to Physical Fitness. L. Golding. Human Kinetics, 1989

APPENDIX B

Review of fitness studies: Single set training versus multiple set training

Introduction

One reason people strength train is to increase muscle strength and hypertrophy (muscle growth). An essential aspect of any strength-training program is the number of sets required for each exercise. Perhaps the most controversial element of any strength-training program is the number of sets required to increase muscular strength and hypertrophy (Carpinelli, 1998). There are two strength-training programs that fuel this controversy.

The first is the single set system or High Intensity Training, which exercises a targeted muscle group for one to three sets. The set is completed when another repetition cannot be properly or safely executed. Performing a lift in this manner is referred to as momentary muscular failure. The single set researchers Westcott [39] and Starkey [29] agree that only one set is necessary to promote strength and hypertrophy.

The second system under debate is the multiple set system. The multiple set researchers Berger [3], Kramer [20], Silvester [28], and Jacobson [17] agree that performing three or more sets is more effective in promoting muscle strength and hypertrophy. Multiple set training is completed with a few sets (normally 3-6) of a given exercise for a prescribed amount of repetitions in each set. Unlike with the single set system, the targeted muscle group is generally not taken to momentary muscular failure. Both of these systems have proven to be effective in increasing strength and muscle hypertrophy.

This literature review provides scientific and empirical evidence comparing the two training systems. Each individual may decide if there is enough evidence to support greater understanding of this research review. To aid in this understanding, a few phrases used in the field of strength and conditioning need more clarification:

- **Muscular failure.** Training to failure is training the muscle to the point where the muscle reaches momentary muscular failure and another repetition is not possible (Stone, 1998).

- **High intensity.** High intensity is generally associated with single set strength training. In high intensity training, each exercise is performed to momentary muscular failure (Westcott, 1995).

- **Strength:** Strength is defined as the ability to produce force (Stone, 1998).

- **One repetition maximum (1-RM).** One repetition maximum is the maximum weight that can be lifted no more than one time with acceptable form (Charette, 1991). Acceptable form means that the primary muscle group used during the exercise lifts the weight without any momentum, bouncing, or jerking motions.

- **Sets.** A set is the number of times the same exercise is performed in a given period of time. Many researchers have studied the effects of multiple sets verses single sets and their relationship to strength and muscle hypertrophy.[you said this in the introduction; do you need to repeat?] In this review, the term "sets" refers to 3 or more.

Strength Studies

Researchers such as Berger, Kremer, Silvester, Jacobson, and Westcott studied the effect of the number of sets on increasing muscle strength. All except Westcott concluded that performing multiple sets is the most effective way to achieve greater muscle strength.

Berger, et al (1962) investigated whether there is an optimal number of a set for increasing strength most rapidly. The subjects of this study were 177 collegiate males experienced in strength training. The subjects were all enrolled in a strength training class at the time of the study. The training program consisted of 3 training sessions per week for 12 weeks. The bench press was selected as the exercise for the study. A 1-RM was tested at the beginning of the study to determine a baseline for each subject. All subjects were tested at the beginning of the program and at 3-week intervals throughout the study. Subjects were randomly divided into 9 groups of 20. Group 1 did 1 set of 2 repetitions (1x2), Group 2 (1x6), Group 3 (1x10), Group 4 (2x2), Group 5 (2x6), Group 6 (2x10), Group 7

(3x2), Group 8 (3x6), Group 9 (3x10). All loads were intended to elicit maximum effort for a given number of repetitions.

The results showed that all groups increased in strength throughout the training period. The strength improvements every 3 weeks were: 1-set groups=7.00 lbs., 2-set groups=6.85 lbs., and 3-set groups=7.95 lbs.

Berger's study indicates that 3-set training increased strength significantly ($p<0.05$) more than training with 1 or 2 sets. Training with 1 set and 2 sets showed similar gains in strength (22.3% and 22.0% respectively), but the groups that trained using 3 sets showed an increase in strength of 25.3%. Berger concluded that training with 3 sets showed the greatest increase in muscular strength.

Like Berger, Kramer et al (1997) investigated which training program produced greater gains in strength. The subjects were 24 competitive collegiate tennis players. Subjects were randomly placed into three groups: two served as the exercise group and the other became the control group. Training sessions were conducted 2-3 times per week for 9 months with identical exercises: bench press, military press, and leg press. The single set group performed 1 set of 8-10 repetitions, whereas the multiple set group performed 2-5 sets of various repetition patterns (3-5, 8-10, and 12-15 RM). Testing was conducted at the beginning of the first, second, third, and ninth months.

Final testing revealed increases in the 1-RM group in the second month for both exercise groups. The results at the end of 9 months determined that the groups using multiple sets achieved greater gains in strength than the groups using a single set. (The strength gains in this study were not revealed.) The control group showed no significant changes.

Silvester et al (1982) undertook to prove that 3 sets of 6 repetitions creates greater increased strength. Four groups of men were tested on biceps strength 3 times per week for 8 weeks. Group 1 performed 1x 8-12 RM, and Group 2 performed 3x6 RM. Both groups performed exercises using barbells. Group 3 performed 1x8 RM, and Group 4 performed 3x6 RM using Nautilus equipment. The strength gains were as follows: Group 1=22%, Group 2=30%, Group 3=25%, and Group 4=19%. Overall, there was no significant difference in any of the training protocols.

Jacobson et al (1986) investigated 2 weight-training programs for knee extensor strength. Two groups performed leg extensions using Nautilus equipment 3 times a week for 10 weeks. Group 1 followed 3x6 protocol, and Group 2 performed a single set to concentric fatigue with forced resistance for the last 3-4 repetitions. Both groups showed similar, yet significant, increases in strength (1=39.2%, 2=31.9%).

Westcott et al (1989) investigated the differences between a single set and 3 sets with 54 men and 23 women over a 10-week period. The subjects trained 3 times a week. They were tested on parallel dips and pull-ups using a Gravitron machine. The results showed similar increases in strength: Group 1 (single set)=4.8% and Group 3 (multiple sets)=5.2%. No data was reported for Group 2.

Strength and Muscle Hypertrophy Studies

Starkey and O'Shea studied the effects of the number of sets on increasing muscle strength and hypertrophy. Both studies concluded that either a single set or multiple sets could be effective in increasing muscle strength and hypertrophy.

Starkey et al (1996) investigated both strength and muscle thickness of the thigh. The subjects followed single set and multiple sets protocol. Thirty-eight males and females were randomly divided into two training groups. The conditioning level of the subjects at the onset of testing was not stated in the study. Each group trained 3 days a week for 14 weeks and were tested on knee extension and knee flexion. Group 1 performed a single set of 8-12 repetitions to muscle failure, and Group 2 performed 3 sets of 8-12 repetitions. Both groups performed the exercises to volitional fatigue.

The findings showed that Group 1 increased knee extension strength by 18.5% and Group 2 had an increase of 13.9%. Knee flexion also showed increases: Group 1=21.8% and Group 2=32.9%. Starkey concluded that both groups increased in strength and thickness and that there were no significant differences between performing a single set or multiple sets, indicating that performing one set is as effective as three sets for increasing muscle strength and thickness.

O'Shea et al (1966) conducted a similar study investigating the effects of a 6-week progressive weight-training program on the development of strength and muscle hypertrophy using the deep knee bend with varying repetitions. The thirty subjects (gender not stated) were randomly divided into 3 groups of 10 for a controlled period of 6 weeks and each group had 3 training sessions per week. Each training session lasted 35 minutes. The subjects were given a 2-week conditioning period to familiarize themselves with the exercise and to avoid the chance of injury. The training programs were as follows: Group 1 performed 3 sets of 9-10 reps, Group 2 performed 3 sets of 5-6 reps, and Group 3 performed 3 sets of 2-3 repetitions. The weight load was increased 5 pounds each week. All groups were tested at 6, 9, and 12 weeks.

The study showed that girth measurements increased 3-6%, and all groups showed considerable improvements during the fourth and sixth weeks. Group 1 increased 4.2%, Group 2 increased 5.2%, and Group 3 increased 3.5%. Strength measurements revealed increases in all groups. The strength increases amounts were not disclosed.

In contrast to the strength studies, these two studies indicate no significant differences in either strength or hypertrophy between single and multiple set strength training.

Muscle Hypertrophy Studies

Charette and Hurley studied the effects of multiple sets on hypertophy in older adults. Their conclusions were similar: multiple sets can produce hypertrophy.

Charette et al (1995) conducted a study to look at muscle hypertrophy response to resistance training in older women. This study showed similar results to the studies described above in regard to the number of sets needed for hypertrophy. Twenty-seven women aged 64-86 were randomly assigned to a control or exercise group. The program lasted 12 weeks and subjects trained 3 days per week. The subjects followed 3 sets of 6-rep protocol. Subjects were tested on 7 resistance exercises for the leg and hip. Testing was conducted during weeks 1, 7, and 12. Muscle strength was initially measured by 1-RM. The control group did not participate in the exercise regimen.

The results showed a significant increase in strength for the exercise group. The control group showed no increase in strength. The final 1-RM values were significantly different between the controls and exercisers for all exercises ($p<0.001$ to $p<0.05$.)

B.F. Hurley et al (1995) investigated the effects of strength training on muscle hypertrophy in older men. The subjects for this study were 35 untrained men aged 50-69. Twenty-three volunteered to participate in the strength-training program, and 12 volunteered to serve as the control group. None of the subjects had been engaged in a regular exercise program during the 6 months prior to this study. Before beginning, a strength assessment, consisting of 4 low-resistance training sessions was administered. Testing was conducted on 6 major muscle groups (4 upper and 2 lower) with a 3-RM benchmark at the beginning and at the completion of the program. The strength-training program consisted of 14 exercises using Keiser variable resistance machines, dumbbells, and floor exercises. Subjects trained 3 days per week for 16 weeks. The training protocol included 3 sets of 15 repetitions and a recovery period of 90 seconds between exercises.

The findings of the 3-RM strength test revealed a 43% increase in strength (p<0.001) and an increase in hypertrophy of 7.2% (p<0.01). There were no significant changes in strength or muscle area in the control group.

Staron et al (1989) investigated the effects of muscle hypertrophy in heavy resistance using 3 sets per exercise. The subjects were 29 women, 14 of whom were considered inactive and sedentary, ranging in age from 19-25 years. The training protocol consisted of a 2-week orientation and pre-conditioning period, followed by an 18-week training period. Following the orientation period, subjects followed 8 weeks of resistance training, 1 week of rest, and 10 weeks of resistance training. The training session consisted of 4 lower body exercises, using 3 sets of 6-8 repetitions with a 2-3 minute recovery between sets. For each participant, the resistance used was based on her 1-RM value. The subjects trained twice per week (Monday and Friday). Testing was performed at the beginning and end of the study. In addition, Wednesday was used once a month for testing.

The results showed that leg girth did not change. However, an increase in lean muscle mass was observed (p<0.05). In addition, subjects showed an increase in lean body mass (p<0.05).

The aforementioned studies indicate that the optimal number of sets for increasing muscle strength and hypertrophy still remains uncertain and controversial. For training effectiveness, performing a single set, two sets, three sets, or more is beneficial for increasing muscle strength and muscle hypertrophy. For training efficiency, however, performing a single set to momentary muscular failure (Westcott and Starkey) can accomplish similar results in less time. Therefore, according to Westcott and Starkey, training the whole body effectively and efficiently, using a single set system, may be more beneficial to the time-conscious fitness enthusiast to achieve strength and hypertrophy within a limited time frame.

When working with athletes or individuals who are involved in other athletic activities, it is wise to train in the most effective and efficient manner. Unfortunately, many coaches and trainers endeavor to replicate the Super Bowl champion's strength-training program without giving thought to the specific needs of the individual or group they are training. Since research shows no significant difference between single and multiple set strength training, the extra time required for multiple set training may be used more efficiently for other activities such as recovery, game preparation, and relaxing—all crucial in developing strength and hypertrophy.

This critical evaluation is not designed to force you to choose which theory is best to increase strength and hypertrophy. Rather, it presents the facts so you may determine which strength program may be optimal. If performing a single set of an exercise to momentary muscular failure can elicit the same results as multiple sets, why waste time in the weight room? On the other hand, if more time in the weight room is enjoyable, and it will achieve similar results, then multiple sets may be more suitable.

CONCLUSION

Burgers and Milkshakes provides the necessary tools to enhance your overall health and fitness level, with variety in developing comprehensive strength and conditioning programs suitable for everyone. *Burgers and Milkshakes* is based on the safest, most efficient, and effective scientifically proven methods available. Everyone is different, with various personalities, goals, fitness levels, and motivation. To unlock your maximal potential in fitness is to incorporate all the principles in this book, be flexible in training programs, and, most importantly, enjoy the process.

BIBLIOGRAPHY

1. American College of Sports Medicine (1995). ACSM's guidelines for exercise testing and prescription.

2. The American Obstetrician and Gynocology Association. 1997 handout. Arauj, David. M.D. (1997). "Expecting questions about exercise and pregnancy," <u>The Physician and</u> <u>Sports Medicine</u>. 25(4) April.

3. Berger, R.A. (1962) "Effects of varied weight training programs on strength," <u>Research Quarterly.</u> 33 329-333.

4. Broderick, J. (1997). Fitness in Today's Times, Inc,—Injury handout and Fitness Assessments.

5. Carpinelli, Ralph, and Robert Otto. "Strength Training: single versus multiple sets," <u>Sports Medicine</u>. 1998 Aug: 26 (2): 73-84.

6. Carron, A.V., H.A. Hausenblas, and D. Mack (1996). "Social influence and exercise: A meta-analysis," <u>Journal of Sport and Exercise Psychology</u>. 18, 1-16.

7. Charette, Susan, Lawrence McEnvoy, Gisela Pyka, Christine Snow-Harper, David Guido, Hurley, B.F., R.A. Redmond, R.E. Partley, M.S. Treuth, M.A. Rodgers, and A.P. Goldberg (1995). "Effects of strength training on muscle hypertrophy and muscle cell disruption in older men," <u>International Journal of Sports Medicine</u>. Vol. 16, No. 6:378-384.

8. Collins, M.A., and T.K. Snow (1993). "Are adaptations to combined endurance and strength training affected by the sequence of training?" <u>Journal of Sports Sciences</u>. 11(6), Dec., 485-491.

9. Chu, Donald (1993). <u>Jumping Into Plyometrics.</u> Human Kinetics, Chicago, Ill.

Darden, Ellington, Ph.D. (1990). <u>The Nautilus Book.</u> Contemporary Books, Chicago, Ill.

10. Dinuble, Nicholas. M.D. (1997). "How to make exercise part of your treatment plan," <u>The Physician and Sports Medicine.</u> 26(7) July.

11. Dishman, R.K. (1987). "Exercise adherence and habitual physical activity," <u>Exercise and Mental Health.</u> 57-83. Hemisphere Publishing Corporation, Washington, D.C.

12. Fahey, T.D. (1997). "Biological markers of overtraining," <u>Biology of Sport.</u> 14(1).

13. Feigenbaum, Matthew, M.D. (1997). "Rationale for Current Guidelines for Adult Fitness Programs," <u>The Physician and Sports Medicine</u>. 25(2) February.

14. Fleck, S. (1982). "The Overtraining Syndrome," <u>National Strength and Conditioning Journal.</u> 4(4):50.

15. Golding, L. (1989). <u>The Y's Way to Physical Fitness.</u> Human Kinetics, Chicago, Ill.

16. Herrick, A.B., and W.J. Stone (1996). "The effects of periodization versus progressive resistance exercise on upper body and lower body strength in women," <u>Journal of Strength and Conditioning Research</u>. 10 (2), May, 72-76.

17. Hurley, B.F. (1995). "Effects of strength training on muscle hypertrophy and muscle cell disruption in older men," <u>International Journal of Sports Medicine.</u> Vol. 16, No. 4: 378-385.

18. Jacobson, B.H. (1986). "A comparison of two progressive weight training techniques on knee extensor strength," <u>Athletic Training.</u> 21 (4): 315-518

19. Katz, J, and B.R. Wilson (1992). "The effects of a six week, low intensity circuit training program on resting blood pressure in females," <u>Journal of Sports Medicine and Physical Fitness.</u> 32(2), Sept.

20. Katz, Warren, M.D. (1998). "Osteoporosis: The role of exercise in optimal movement," <u>The Physician and Sports Medicine</u>. 26(2) February.

21. Kramer, J.B., and M.O. Stone (1997). "Effects of single vs. multiple sets of weight training," Journal of Strength and Conditioning Research. 11 (3) Aug., 143-147.

22. Mcardle, W., F. Katch, and V. Katch (1996). Exercise Physiology, 4th Edition. Williams & Wilkins, Baltimore, Md.
    Mannie, K. (1997). "Five Major Facts on Player Development,"Coach and Athletic Director. 66 (6).

23. Marib, E. (1991). Human Anatomy and Physiology, 3rd Edition. Benjaman and Cumings, Redwood City, Cal.

24. Miller, J.A. (1983). "Beginning weight training," Strength and Health. 76:17.

25. National Sports Performance Association, (1997). Certified Conditioning Specialist Training Manual.

26. O'Shea, Patrick (1966). "Effects of selected weight training programs on the development of strength and hypertrophy," Research Quarterly. 37:95-102.

27. Riley, D. (1998). Washington Redskins Conditioning Manual.

28. Sallis, J.F., and M.F. and Hovell. (1990). "Determinants of exercise behavior,". Exercise and Sport Science Reviews. 18:307-330. Williams & Wilkins, Baltimore, Md.

29. Silva, J. M., III. (1990). "An analysis of the training stress syndrome in competitive athletics," Journal of Applied Sport Psychology. 2, 5-20.

30. Silvester, L.J. (1982). "The effects of variable resistance and free-weight training programs on strength and vertical jump," National Strength and Conditioning Association Journal. 3 (6):30-33.

31. Simkin, L.R., and A.M. Gross (1994). "Assessment of coping with high-risk situations for exercise relapse among healthy women," Health Psychology. 13, 274-277.

32. Starkey, D.B., M.L. Pollock, and Y. Ishda (1996). "Effect of resistance training volume on strength and muscular thickness," Medicine and Science in Sports and Exercise. 28(10) 1311-1320.

33. Staron, R.S., E.S. Malicky, M.J. Leonardi, J.E. Falkel, F.C. Hagerman, and G.A. Dudley (1989). "Muscle hypertrophy and fast fiber type conversions in heavy resistance-trained women," European Journal of Applied Physiology. 60:71-79.

34. Stone, M.H. (1990). "Muscle conditioning and muscle injuries," Medicine and Science in Sports and Exercise. 22(4), Aug., 457-462.

35. Stone, M.H. (1987). Weight Training: A Scientific Approach. Burgess International, Minneapolis, Minn.

36. Thompson, L.V. (1994). "Effects of age and training on skeletal muscular physiology and performance," Physiology and Therapy. 74:71-81.

37. Maiorca, S. and D. Martin (1995). United States Bobsled and Skeleton Federation Conditioning Manual.

38. Wescott, Wayne (1987). Physiological Principles and Training Techniques. Allan & Balon, Inc., Newton, Mass.

39. Westcott, Wayne. (1992). "A new look at youth fitness," American Fitness Quarterly. Vol. 11.

40. Westcott, Wayne, . (1996). Building strength and stamina. Human Kinetics, Chicago, Ill.

41. Westcott, Wayne, and K. Greenberger (1989). "Strength training research: sets and repetitions," Scholastic Coach. 58: 98-100

42. Wilkius, K.E. (1980). "The uniqueness of the young athlete: musculoskeletal injuries,"
American Journal of Sports Medicine. 8:377-82.

43. Willoughby, D.S. (1992). "A comparison of three selected weight training programs on the upper and lower body strength of trained males," Applied Research in Coaching and Athletes Annual. 124-146

ABOUT THE AUTHOR

David provides individuals and companies with simple steps and knowledge that may improve their personal and professional growth and health. He has worked with the FDA, ExxonMobil, Champion Jog Bra, and many other companies and schools to promote corporate wellness and improve individual's quality of life. He draws from an extensive background in physical education and sports management from Skidmore College.

He was an Assistant Strength and Conditioning Coach for the United States Olympic Bobsled Team and the Baltimore Ravens. He also was a fitness consultant for the Joint Chief of the United States Army. He owned a personal training company for 8 years and taught thousands of aspiring personal trainers as a course instructor with the National Strength Professionals Association.

Currently, when David is not traveling and proving wellness programs to companies, he spends active time with family and friends in Olney, MD.

978-0-595-34776-6
0-595-34776-2